PROFESSIONAL STANDARDS FOR PREPARING, HANDLING, AND USING EXPLOSIVES

THOMAS MORDECAI

PROFESSIONAL STANDARDS FOR PREPARING, HANDLING, AND USING EXPLOSIVES

PALADIN PRESS
BOULDER, COLORADO

Professional Standards for Preparing, Handling, and Using Explosives
by Thomas Mordecai

ISBN 0-87364-807-2
Printed in the United States of America

Published by Paladin Press, a division of
Paladin Enterprises, Inc., P.O. Box 1307,
Boulder, Colorado 80306, USA.
(303) 443-7250

Direct inquiries and/or orders to the above address.

CONTENTS

WARNING

The procedures to be followed in this manual and the resulting end products are extremely dangerous. Whenever dealing with explosive materials, safety standards should be followed precisely. Failure to do so could result in serious injury or loss of life.

It is your responsibility to check and obey all relevant federal, state, and local laws concerning the possession, handling, storage, transportation, or use of explosive materials. Ignorance of the law is no excuse.

The author and the publisher disclaim any liability from any damages or injuries of any type that a reader or user of information contained herein may encounter. This material is for *informational purposes only*.

INTRODUCTION

The guidelines herein reflect current industry-standard recommendations regarding the safe storage, handling, and use of commercial explosive materials. Knowledge of these guidelines is an absolute must for the occasional explosives user, potential user, interested nonuser, and investigative law enforcement or security operative who may come across such materials in the course of his or her duties.

As all books and videos on related subjects point out, a failure to follow such guidelines can easily result in injury or death. Until now, however, unless you were already an experienced or professional user of explosive materials, it may have proven difficult to find out just what these guidelines were. This book rectifies that problem and brings relevant safety and technical information together in the form of a single handy reference.

Aside from saving your life, knowledge of the guidelines and terms explained in this book will enable you to hold your own in any conversation with professional explosives users or materials suppliers, so you can come across as an expert, not an experimenter! If you hope to find employment in the field of commercial blasting, this book will give you a head start.

An easy-to-read "Do's and Don'ts" format has been adopted for the guidelines section, and a comprehensive glossary of explosives terms is included. This information will prove an invaluable reference source in its own right.

CHAPTER 1

HANDLING EXPLOSIVE MATERIALS

GENERAL

DO ensure that all persons engaged in general explosive handling operations are at least 18 years old, mentally and physically capable of performing the work, and familiar with the characteristics and hazards of the materials they are to handle.

DO ensure that such persons are able to comprehend spoken and written instructions in the English language.

DO ensure that such persons are not addicted to alcohol or drugs.

DO ensure that personnel in charge of magazines are at least 21 years old.

DO ensure that personnel assigned as blasters are at least 21 years old.

DO ensure that personnel assigned specific explosives-related tasks are completely familiar with all aspects of the work involved and possess all required local, state, and federal permits and licenses.

DO keep explosive materials away from unauthorized persons, children, and animals.

DO use only permissible explosive materials in gassy, flammable, or dusty atmospheres.

DO close properly the packaging of partially used explosive materials.

DON'T remove explosive material from packets or housings intended to be left on during use.

DON'T use any explosive material unless you are completely familiar with its properties, effects, and applicable safe-use procedures, or unless you are acting under the direct instruction of someone who does.

DON'T carry any type of explosive material in your clothing, i.e., in your pockets.

DON'T handle explosive materials during an electrical storm.

DON'T try to fight fires involving explosive materials. Move yourself and others to a safe distance and call the emergency services.

DON'T allow metal slitters or other metal tools to touch the metal fastenings on explosive material packaging during opening.

DON'T mix or try to alter the contents of explosive materials.

DON'T allow explosive materials to come into direct contact with skin, eyes, or foodstuffs.

DON'T breathe fumes, vapors, or dust from explosive materials.

CHAPTER 2

TRANSPORTING EXPLOSIVE MATERIALS

GENERAL

Updates of regulations concerning the commercial conveyance of explosive materials over public highways are to be found in Department of Transportation (DOT) 49 CFR. Specific state, city, and local laws exist and must be obeyed.

DO ensure that the drivers of trucks engaged in commercial explosives transport operations are at least 21 years of age.

DO ensure that the vehicle is in good mechanical condition and strong enough to carry the required load.

DO drive at a sensible speed and avoid hard braking and cornering.

DON'T ever leave a vehicle carrying explosives unattended.

DON'T park vehicles carrying explosives in congested areas or close to people.

LOAD SAFETY

DO keep explosive materials in their original containers or in day boxes when transporting them in enclosed vehicles.

DO ensure that explosive materials are kept in closed containers fastened securely to the truck bed when transporting them in open-bodied vehicles.

DO load, pack, and unload explosive materials carefully.

DON'T carry detonators or other explosive materials in the glove compartments of vehicles.

FIRE SAFETY

DO carry at least two dry chemical fire extinguishers (multipurpose type).

DO use the correct gear to supplement braking and reduce brake and wheel temperatures during steep or long gradients.

DO park out of direct sunlight during rest breaks.

DON'T try to fight a cargo fire. Separate tractor or tow vehicle from the trailer if possible and drive it at least 200 feet away. Stop traffic in both directions. Assign someone to call the fire department and police and explain the situation and the nature of the load. Evacuate occupants of nearby buildings. Keep everyone at least 200 feet away from the vehicle.

TIRE SAFETY

DO check tire pressure and condition before the journey and at regular intervals during it.

DO check wheel lugs for tightness before the journey and at regular intervals during it.

DO remove any tires that become damaged or overheated during the journey.

DO use water, dry chemical, or dirt on an overheated tire or small tire fire in an attempt to cool it or douse it before abandoning the vehicle and initiating emergency procedures.

DO remember that proven negligence is a criminal offense punishable by law.

DON'T allow lighters, matches, or any other source of ignition within 50 feet of a parked explosives-laden vehicle.

DON'T place tires that have been removed because of overheating in positions in the vehicle that could start a fire.

CHAPTER 3

STORING EXPLOSIVE MATERIALS

GENERAL

DO store explosive materials of any type in accordance with specific federal, state, and local laws. Doing so will help ensure your own safety and that of the general public and help control the illegal distribution and sale of explosive materials. Explosive materials should be stored in specifically built magazines located and constructed so as to prevent unauthorized access, limit the deterioration of the materials by protecting them from the effects of weather, and protect them from fire and strikes by bullets. Magazines should display a clear indication of their contents, e.g., "*Danger: Explosives—Keep Away.*"

STORAGE SECURITY

DO keep the magazine locked at all times.
DO check frequently that the magazine is secure and undamaged (at least every seven days).
DO immediately protect explosive materials from further theft if your magazine is broken into.
DO immediately repair the magazine.
DO call BATF (1-800-800-3855) immediately if you lose or have explosive materials stolen.

STORAGE SAFETY

DO ensure that magazines are separated from other magazines, roads, railways, and inhabited buildings as per the American Table of Distances extract given below:

AMERICAN TABLE OF DISTANCES

QEM	=	Quantity of explosive materials (in pounds)
MT	=	More than
NMT	=	Not more than
IB	=	Inhabited buildings (minimum distance to)
PH3000	=	Public highways with a traffic volume of less than 3,000 vehicles per day (distance to)
PHPR	=	Public highways with a traffic volume greater than 3,000 vehicles per day and passenger railways (distance to).
MAG	=	Magazine/s (distance to). If multiple magazines are co-located at distances less than the suggested minimum, they must be considered a single magazine and the total QEM therein used to determine other safe distances.

QEM	MT	0	NMT	5	IB=140	PH3,000=60	PHPR=102	MAG=12
QEM	MT	5	NMT	10	IB=180	PH3,000=60	PHPR=128	MAG=16
QEM	MT	10	NMT	20	IB=220	PH3,000=90	PHPR=162	MAG=20
QEM	MT	20	NMT	30	IB=250	PH3,000=100	PHPR=186	MAG=22
QEM	MT	30	NMT	40	IB=280	PH3,000=110	PHPR=206	MAG=24
QEM	MT	40	NMT	50	IB=300	PH3,000=120	PHPR=220	MAG=28
QEM	MT	50	NMT	75	IB=340	PH3,000=140	PHPR=254	MAG=30
QEM	MT	75	NMT	100	IB=380	PH3,000=150	PHPR=278	MAG=32

Distances are given in feet and assume an *unbarricaded* structure. For a *barricaded* structure, halve the distances given. (See the glossary for definitions of barricaded and unbarricaded structures.)

DO ensure that the magazine interior is kept clear, dry, cool, and ventilated.

DO use stocks of explosives of a first-in, first-out basis, i.e., use the older stocks first.

DO check shelf-life dates to confirm that the material remains safe to use.

DO store explosive materials only in their original packaging.

DON'T undertake any activity inside the magazine aside from the inspection or removal of explosive materials.

DON'T smoke or allow lighted cigarettes, matches, lighters, or any source of ignition within 50 feet of the magazine.

DON'T let combustible material accumulate within 25 feet of the magazine.

DON'T attempt to repair any part of the magazine internally or externally with the explosive material still inside.

DON'T store detonators with other explosive materials.

MINIMUM SEPARATION DISTANCES

DO maintain the minimum separation distance of ammonium nitrate and blasting agents (acceptors) from explosives and blasting agents (donors), as shown below:

TABLE OF RECOMMENDED SEPARATION DISTANCES OF AMMONIUM NITRATE AND BLASTING AGENTS FROM EXPLOSIVES OR BLASTING AGENTS

Donor Weight		Minimum Separation Distance of Acceptor When Barricaded (in Feet)		
Pounds Over	Pounds Not Over	Ammonium Nitrate	Blasting Agent	Minimum Thickness of Artifical Barricades (in Inches)
	100	3	11	12
100	300	4	14	12
300	600	5	18	12
600	1,000	6	22	12
1,000	1,600	7	25	12
1,600	2,000	8	29	12
2,000	3,000	9	32	15
3,000	4,000	10	36	15
4,000	6,000	11	40	15
6,000	8,000	12	43	20
8,000	10,000	13	47	20
10,000	12,000	14	50	20
12,000	16,000	15	54	25
16,000	20,000	16	58	25
20,000	25,000	18	65	25
25,000	30,000	19	68	30
30,000	35,000	20	72	30
35,000	40,000	21	76	30

13

40,000	45,000	22	79	35
45,000	50,000	23	83	35
50,000	55,000	24	86	35
55,000	60,000	25	90	35
60,000	70,000	26	94	40
70,000	80,000	28	101	40
80,000	90,000	30	108	40
90,000	100,000	32	115	40
100,000	120,000	34	122	50
120,000	140,000	37	133	50
140,000	160,000	40	144	50
160,000	180,000	44	158	50

If the ammonium nitrate or blasting agent is *not* barricaded, the distances given should be multiplied by a factor of six.

When storage is in a bullet-resistant magazine recommended for explosives or when the storage facility is protected by a bullet-resistant wall, the distances and barricade thicknesses given in the American Table of Distances apply.

MAGAZINE FIRES

DON'T fight a fire inside a magazine.

DO keep the numbers of the fire and police departments in a conspicuous place outside the magazine.

DO call them immediately in the event of a fire and tell them the nature of the magazine contents.

DO evacuate personnel out of the area of the magazine to a radius of at least 2,500 feet.

DO organize a perimeter patrol to keep people from reentering the area.

DO help authorized personnel fight a fire *outside* the magazine.

CHAPTER 4

PREPARING, LOADING, AND USING EXPLOSIVE MATERIALS

GENERAL

DO check boreholes before loading to confirm that they are safe and free from unfired explosives.

DO take precautions to prevent the buildup of static electricity when using pneumatic loading.

DO check surfaces and faces for unfired explosive materials before drilling boreholes.

DO keep your body behind a borehole when loading, stemming, or tamping.

DO design a blast to minimize flyrock, ground vibration, and air blast.

DO clear, the blast area of all personnel, vehicles, equipment, and explosive materials before firing.

DO post guards around the safety perimeter before firing.

DO use a blasting mat when blasting near any locations at risk from the effects of flyrock.

DO initiate all explosions from a location well away from the blast area and flyrock danger area.

DO remain in the safe area until all postexplosion debris, smoke, and fumes have settled.

DON'T have more explosive material than is needed for immediate use stacked in the working area during loading.

DON'T redrill a blasthole that has contained explosive materials.

DON'T drill directly into explosive materials.

DON'T start a borehole in a bootleg.

DON'T force explosive materials into a borehole.

DON'T drop cartridges directly onto primer charges.

DON'T load boreholes containing burning material or boreholes in material having a temperature greater than 150°F.

DON'T use springing near other boreholes that are already loaded with explosive material.

DON'T use excessive force when tamping.

DON'T tamp an explosive material that has been removed from a cartridge housing designed to be left on during use.

DON'T tamp explosive material with improvised metal devices. The only metal device that can be safely used is a jointed nonsparking metal pole with non-ferrous metal connectors.

DON'T damage or kink detonator wires, safety fuse, detonating cord, plastic tubing, or shock tubing when loading or tamping.

DON'T assemble primers inside magazines or near other large quantities of explosives.

DON'T assemble more primers than are required for immediate use.

DON'T allow any source of ignition within 50 feet of a blast site.

DON'T ever fire a charge without first having received an OK signal from the individual in charge.

DON'T ever fire a charge without first sounding an adequate warning signal.

DON'T insert anything into a detonator designed for initiation by safety fuse, except safety fuse.

DON'T allow shooting or loaded firearms in the vicinity of explosive materials, magazines, or explosive-carrying vehicles.

DON'T allow explosive materials to be exposed to any sources of heat exceeding 150°F or to flame or any other source of ignition, except where such exposure is part of the approved initiation process.

DON'T allow explosive materials to be struck, impacted, or compressed by any object, except where such techniques are part of the approved loading sequence and the forces involved are within safe, specified parameters.

DON'T subject explosives to excessive friction.

DON'T try to disassemble detonators.

DON'T pull detonating cord, wires, shock tubing, safety fuse, or any other initiating system out of a detonator or delay device.

DON'T use explosive materials that have been water-soaked, even if they appear to have dried out thoroughly.

DON'T allow persons to handle, or be in close proximity to other explosive materials when a charge is fired.

DON'T fire a charge from an area within close proximity to other explosive materials.

PREPARING PRIMERS

DO make holes in explosive material only with a non-sparking punch designed for the job.

DO have the detonator pointing in the direction of the main charge.

DO ensure that the detonator is inserted fully into the primer cartridge and does not protrude from it.

DO secure the detonator in the primer charge in a manner that ensures no tension is placed on the detonator wires, safety fuse, detonating cord, or plastic tubing.

DON'T try to force a detonator directly into explosive material.

DON'T use a punch on explosive material that is extremely hard or frozen solid.

DON'T try to enlarge a hole in a cast primer or booster to accept a detonator.

DON'T try to use a cast primer or booster if the detonator hole is too small.

PREPARING PRIMERS WITH ELECTRIC DETONATORS

DO follow the sequence outlined below when preparing primers with electric detonators and cartridges less than 4 inches in diameter:

A. Punch a straight hole into the center of the cartridge.

B. Insert the detonator.

C. Tie-off the detonator legwires with a half-hitch knot, taking care not to pull the wires too tightly.

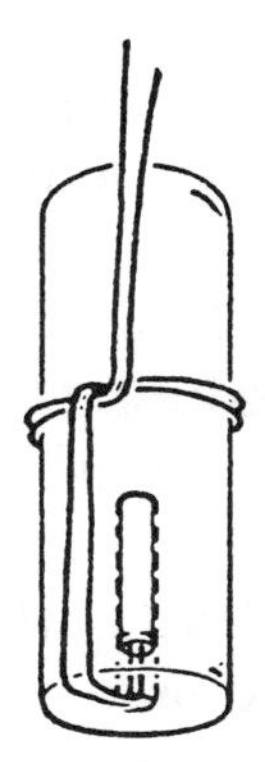

Suggested method of assembling primer with cartridge diameter of less than 4 inches and electric detonator.

DO follow the sequence outlined below when preparing primers with electric detonators and cartridges of 4 inches or more in diameter:

A. Punch a slanting hole from the center of one end of the cartridge through its side some 2 or more inches from the other end.

B. Fold the legwires sharply about 12 inches from the detonator.

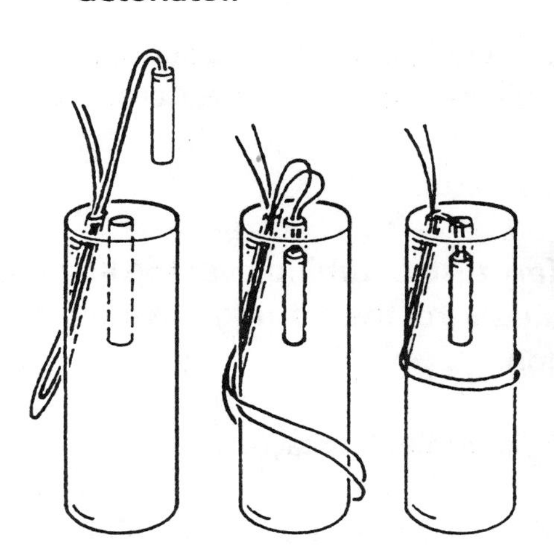

Suggested technique of assembling primer charge using cartridge of 4 inches diameter or more and an electric detonator.

C. Push the folded legwires in through the hole at the end of the cartridge and out through the hole in the side.

D. Open out the folded wires and pass the loop thus formed over the nonholed end of the cartridge.

E. Punch another hole straight into the end of the cartridge next to the first. Insert the detonator into this hole and then take up the slack remaining in the wires.

PREPARING PRIMERS WITH NONELECTRIC DETONATORS

DO follow one of the sequences outlined below when assembling primers with nonelectric detonators.

REVERSE PRIMING

A. Punch a straight hole into one end of the cartridge. The hole should be longer than the detonator.

B. Insert the detonator.

C. Fold the safety fuse or tubing over the holed end and down the side of the cartridge, as indicated in the illustration.

D. Tape the fuse or tubing in place.

NOTE: This technique can be used with miniaturized

detonating cord but *only if* the explosive to be used is insensitive to initiation by the miniaturized cord directly.

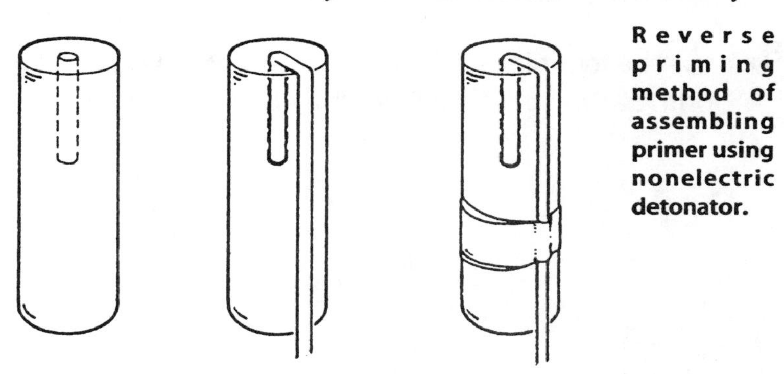

Reverse priming method of assembling primer using nonelectric detonator.

SIDE PRIMING

A. Punch a slanting hole into the side of the cartridge. The hole should be longer than the detonator.

B. Insert the detonator.

C. Secure the safety fuse or tubing with tape where it exits the cartridge, as shown.

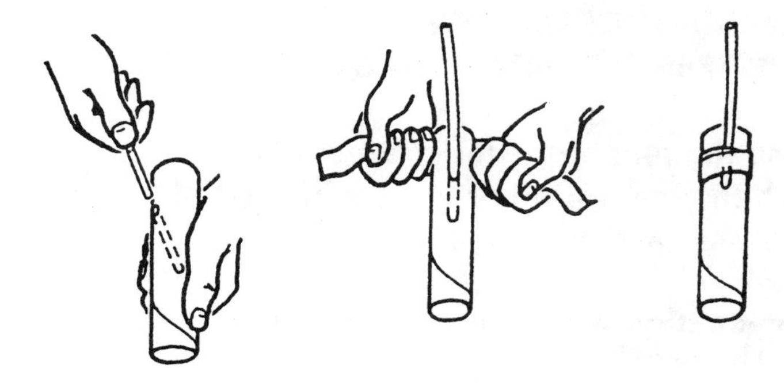

Side priming method of assembling primers using nonelectric detonator.

ASSEMBLING PRIMERS WITH CAST BOOSTERS AND ELECTRIC DETONATORS

DO use the technique illustrated below when assembling primers with cast boosters and electric detonators.

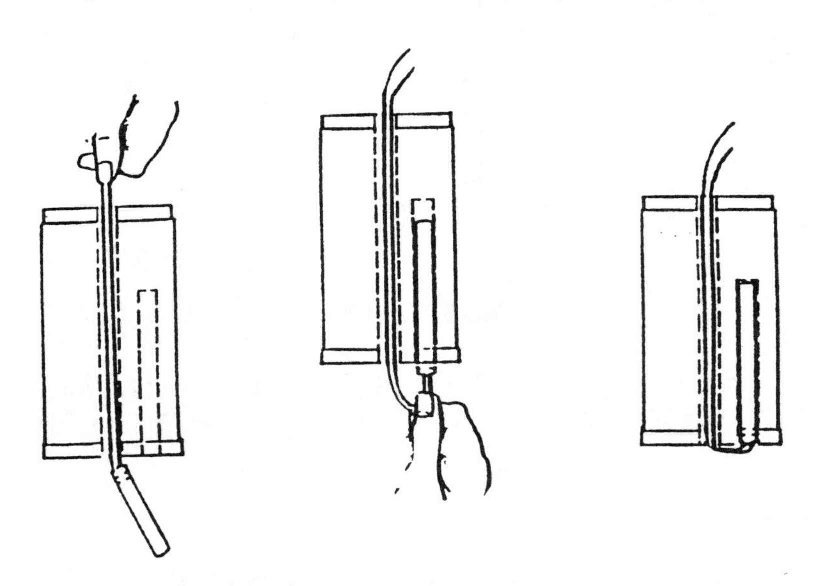

Suggested method of assembling primer with cast booster cartridge and electric detonator.

ASSEMBLING PRIMERS WITH CAST BOOSTERS AND DETONATING CORD

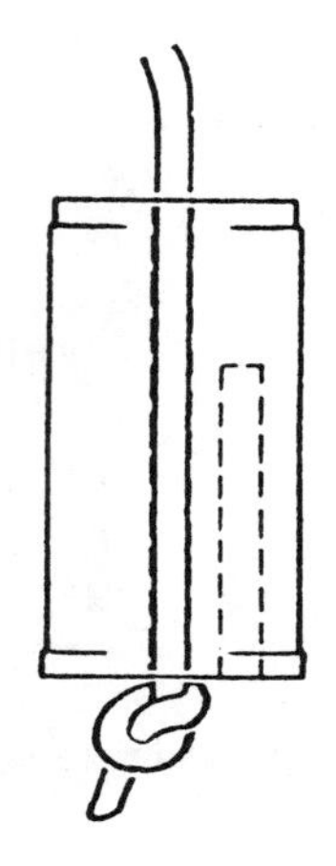

DO use the technique illustrated below when assembling primers with cast boosters and detonating cord.

Suggested method for assembling primer with cast booster and detonating cord.

ASSEMBLING PRIMERS WITH PLASTIC FILM CARTRIDGES AND NONELECTRIC DETONATORS

DO use the technique illustrated below when assembling primers with plastic film cartridges and nonelectric detonators.

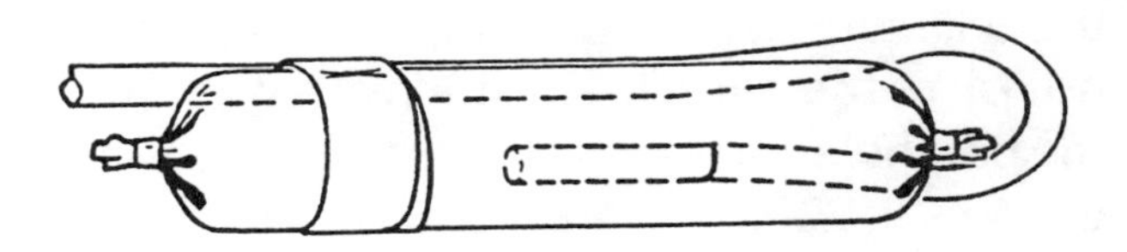

Suggested method for assembling primer with plastic film cartridges and nonelectric detonators.

ASSEMBLING PRIMERS WITH PLASTIC FILM CARTRIDGES AND ELECTRIC DETONATORS

DO use the method shown below when assembling primers with plastic film cartridges and electric detonators.

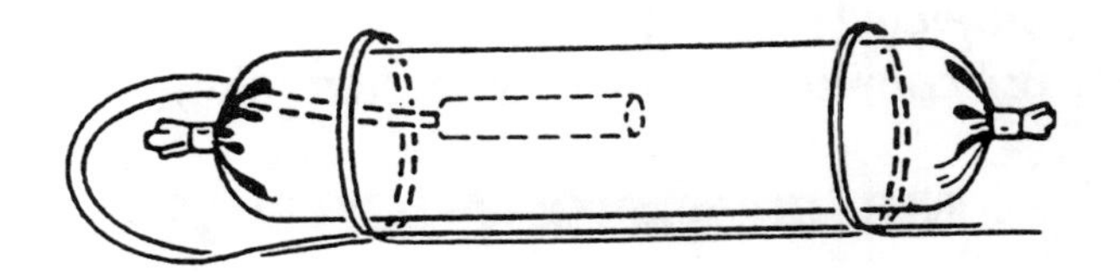

Suggested method of assembling primers with plastic film cartridges and electric detonators.

STEMMING

Any hole containing explosive material to be fired should be stemmed. The confinement provided by stemming allows the explosive material to realize its maximum potential for work and serves to lessen the effects of air blast and flyrock. In the case of the rotational firing of closely located boreholes, the stemming helps prevent the as-yet unfired charges being sucked out of the borehole.

DO use drill-cutting dust for stemming only when there is no risk of the particles mixing with water present in the borehole and being rendered ineffective.

DO use .25- to .5-inch clean, crushed stone for stemming in large-diameter boreholes that contain significant amounts of water. Clean, sharp sand can be used in small-diameter holes.

DO use stemming material packed in tamping bags or flexible plastic tubes when stemming horizontal holes.

DO be careful not to damage legwires, detonating cord, fuse, or other firing system parts when stemming the hole.

DON'T use large stones for stemming. They can jam in the hole and prevent proper confinement or be propelled from the hole when it is fired, much like a projectile from a gun.

DON'T use combustible material for stemming.

INITIATING EXPLOSIVE MATERIALS

Nonelectric Initiation

Safety Fuse and Nonelectric Detonator

DO undertake a test burn of safety fuse to establish its burning rate before use.

DO handle fuse carefully so as not to damage the outer covering.

DO follow the method outlined below when assembling a blasting cap and safety fuse:

A. Cut off an inch or so to ensure a completely dry end.

B. Measure the required length of fuse and cut squarely with a fuse cutter. *Do not use a knife.*

C. Examine the open end of the blasting cap for debris or moisture. Attempt to remove any such foreign matter *only* by inverting the blasting cap. If foreign matter cannot be removed in this manner, do *not* use the cap.

D. Hold the end of the safety fuse vertically and gently install the cap over it.

E. Crimp the cap no further than .25 inch from the point where the fuse enters, using only commercial cap crimpers.

DON'T twist or screw the fuse into the detonator.
DON'T use your teeth or anything other than commercial crimpers for crimping.
DON'T use safety fuse lengths of less than 3 feet.
DON'T cut the fuse until you are ready to install it in the detonator.
DON'T try to remove a detonator from safety fuse once it has been crimped in place.

Lighting Safety Fuse

DO ensure that the fuse length is sufficient to let you reach a safe location before initiation.
DO ensure, before lighting, that the charge is covered

with enough stemming to protect it from sparks, spits, and fuse heat.

DO have another person with you when you light the fuse.

DON'T light safety fuse with anything other than a commercial lighter system. For single-fuse ignition, use thermalite connectors, pull-wire lighters, or hot-wire lighters. For multifuse ignition, use ignitor cord and thermalite connectors.

DON'T hold explosive charges when lighting safety fuse.

DON'T use safety fuse in agricultural blasting operations.

DON'T load a borehole with a charge or primer that has a lighted fuse.

Initiating Detonating Cord

DO use the technique below when initiating detonating cord with nonelectric detonators fired by safety fuse.

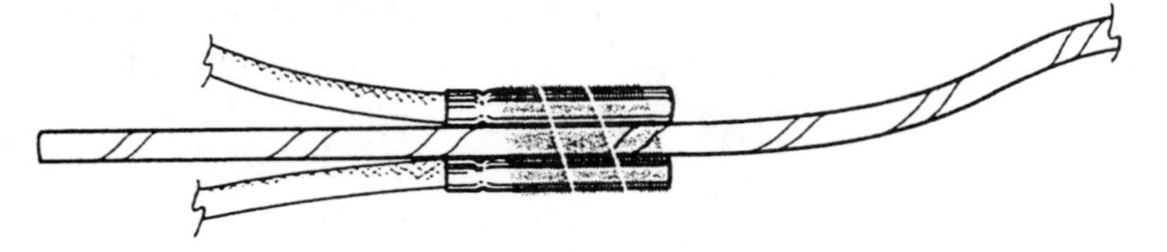

Suggested method of initiating detonating cord with nonelectric detonators.

DO use the technique illustrated below when initiating detonating cord with electric detonators or nonelectric detonators fired by means *other* than safety fuse.

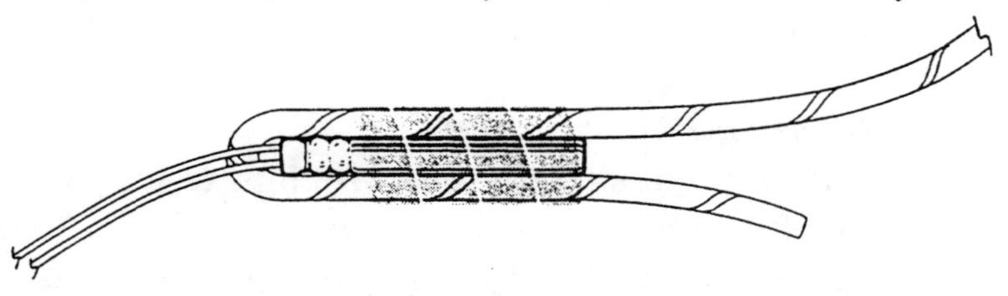

Suggested method of initiating detonating cord with electric detonators or nonelectric detonators fired by means other than safety fuse.

DO ensure that the detonator points in the direction of initiation.

DO treat detonating cord with the same respect as other explosives.

DO affix detonators to detonating cord with tape or purpose-made clips.

DO cut the detonating cord from the roll before loading.

DO attach the initiating detonator at least 6 inches from the end of the detonating cord.

DO use a suitable booster to initiate wet detonating cord.

DO use only a sharp knife or blade to cut detonating cord.

DON'T cut detonating cord with scissors or similar scissor-action tools.

DON'T use detonating cord that is damaged.

DON'T attach initiating detonators until the blast area has been cleared and secured.

Electrical Initiation

DO check wire ends for cleanliness before making connections.

DO ensure that the firing circuit is insulated from the ground and other potential conductors.

DO use only firing currents of a magnitude specified by the detonator manufacturer.

DO check the firing circuit for continuity and resistance using only approved blasting meters.

DO keep detonator leg and connecting wires shunted and isolated from the firing current source until ready for firing.

DON'T mix detonators from different manufacturers in the same circuit.

DON'T mix detonators of different types, although from the same manufacturer, unless such mixing is specified as being safe by the manufacturer in question.

DON'T try to test for circuit continuity with any system other than an approved blasting galvanometer or test meter. If the test item doesn't specify "blasters," DON'T use it.

DON'T use aluminum wiring in a blasting circuit.

DON'T make the final connection to the firing current source until you are sure all personnel are clear of the blast site.

Shock Tube Initiation

DO cut and splice shock tube units only in accordance with the manufacturer's instructions.

DO ensure that detonating cord/shock tube connections are at right angles.

DO run the shock tube to the borehole in a straight line.

DO keep the shock tube taut.

DON'T tie two lengths of shock tube together.

DON'T allow vehicles to drive over the shock tube.

Gas-Initiated Systems

DO use specially designed boosters or tube protectors.

DO avoid the blast area after connections are readied for firing unless the complete system has been purged and disconnected from the primary ignition source.

Miniaturized Detonating Cord

DO use only explosives that are *insensitive* to initiation via miniaturized detonating cord.

DON'T try to join two lengths of miniaturized detonating cord.

29
MISFIRES

Misfires: Safety Fuse and Nonelectric Detonators

DO cordon off the blast area and wait at least 30 minutes after the misfire before reentering.

Misfires: Electrical Detonators and Nonelectric Detonators Other Than Those Initiated with Safety Fuse

DO cordon off the blast area and wait at least 15 minutes after the misfire (unless a different time is suggested by the manufacturer of the system in question) before reentering.

DO disconnect and shunt the lead line and secure the blasting machine.

DO check electrical circuits for continuity with a blasting ohmmeter.

Misfires: General

DO carefully inspect the area for obvious firing system breaks and cutoffs, and assess the situation carefully before deciding what specific remedial action to take.

DO use detonation (by repriming and refiring) to dispose of misfired explosive material if there is sufficient cover or burden to contain the blast.

DO use only a plastic or rubber airline or hose pipe, or a plastic or wooden tamping pole to remove stemming from blast holes.

DO use a replacement primer of high strength to refire any charge(s) that may have been wetted during the removal of stemming.

DO wash out—or if the ground around it is broken, lift out—the charge and remove to a magazine for subsequent disposal only if the misfired charge cannot be refired.

DO be aware that flyrock presents a great danger when refiring misfired holes. Flyrock should always be expected from misfires in which a portion of the blast has fired. Mats or some other type of screening should be employed if required.

DO always investigate misfires to determine their cause and thereby prevent their repetition. Typical causes of misfires are incorrectly made primers, damage to the fuse powder train, improper electrical or physical connections, damage caused to detonator legwires or fuse powder train during loading, the use of non-water-resistant materials in wet areas, and incorrect loading techniques.

DON'T return to the blast area for at least 60 minutes after the refiring of a misfired charge.

DON'T attempt to pick out, bore, or drill out explosive materials that have misfired. Only experienced, competent personnel fully aware of the blast design, location, and type of explosive materials in question should deal with the misfire.

Partial Misfires

Partial misfires often result from cutoff holes, the use of deteriorated explosive materials, the effect of water on the explosive materials, incorrect or inadequate priming, and incorrect loading techniques or cuttings causing cartridges to become isolated. Borehole cutoffs can be minimized by priming the explosive correctly at the bottom of the hole; by designing the round with proper consideration to spacing, burden, and visible partings or seams; and by properly delaying the blast.

HANGFIRES

Hangfires occur when part or all of the explosive material in a borehole begins to burn; this burning can often result in an explosion. A possible cause of burning explosive material is the arcing of a delay electrical detonator. Arcing can be eliminated by using energy-limiting firing switches or capacitor discharge blasting machines. A hangfire might also be caused by an interruption of the detonation wave in an explosive material by (for example) drill cuttings, or the ignition of misfired material via heat from the detonation of material in adjacent or nearby holes.

PROTECTION AGAINST RADIO FREQUENCY HAZARDS (RADHAZ)

Overview

In the presence of a strong RF field, electric detonator legwires or firing circuit wires will act as an antenna system and the RF field will induce a current to flow within that system. This happens whether or not the legwires or circuits are connected to the blasting machine and regardless of whether the legwires or circuit wires are shunted or open. The amount of energy actually reaching the detonator, however, will depend on many factors, including the specific configuration and orientation of the firing circuit wires and the strength and frequency of the RF field itself.

In the case of transportation in vehicles equipped with two-way radio, no hazard will exist if electrical detonators are kept in their original containers or metal boxes with a close-fitting metal lid with the legwires coiled and folded. This configuration is extremely effective at protecting against RF induction. Transmitters must not be used when electrical detonators are being removed or replaced in their packing boxes. As protection against friction and shock, metal boxes used for transporting electrical detonators should be lined with a soft material such as foam rubber or wood.

The most dangerous situation exists when the electrical detonator is configured into a firing circuit, specifically when the legwires or attached circuit wires are raised several feet off the ground and the length of the wiring is such that it is resonant at half of the RF wavelength or some multiple of it, and the electric detonator is located at a point where induced RF current is at a maximum. In the example below, the electrical firing circuit is forming what, in radio terms, would be called a "center-fed" dipole antenna.

"Dipole Antenna" RF Pickup Circuit

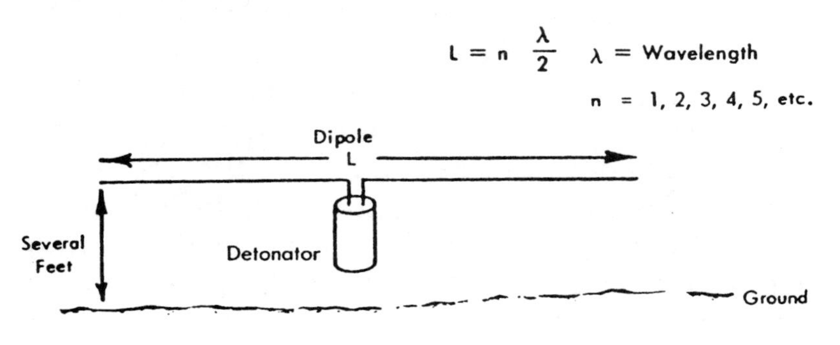

1/2-wave dipole.

An equally hazardous circuit exists when the electric detonator is attached to one end of the wiring, the wiring is raised several feet off the ground, the length of the wiring is equal to one-quarter of the RF wavelength or an odd multiple of it, and the wiring is grounded via the detonator. This forms a "long-wire" antenna.

"Long-Wire Antenna" RF Pickup Circuit

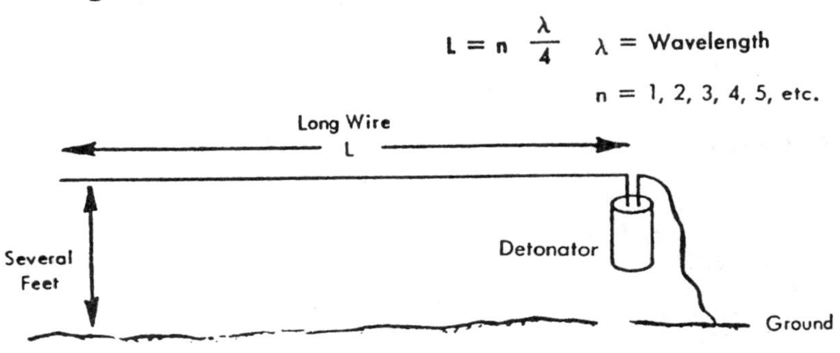

Long-wire antenna.

Both the above antennas will prove most efficient at collecting RF when they are *parallel* to a *horizontal* transmitting antenna or *pointed toward* a *vertical* antenna.

Loops in an electrical firing circuit are also susceptible to RF pickup. The loop is sensitive to the magnetic component of the RF signal, and maximum pickup occurs when the loop is in the same plane as the transmitting antenna. A rule of thumb is that the larger the loop, the greater its capability in capturing RF signals.

The illustration below shows the preferred and "worse case" layouts for transmitting antennas and electrical firing circuits into which a loop might be introduced.

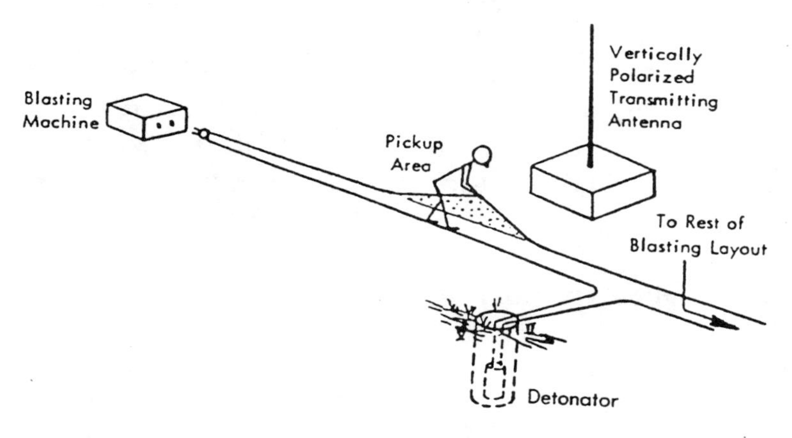

Preferred loop.

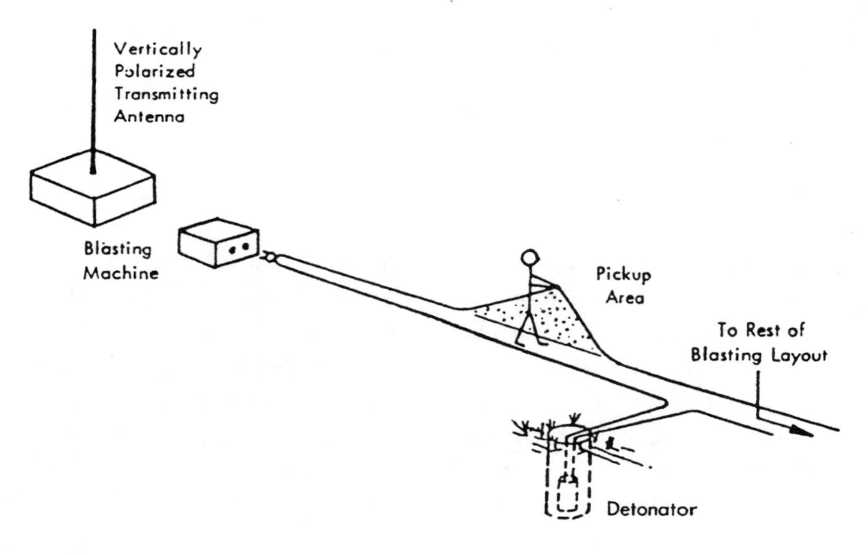

Worst-case loop.

It is worth noting that despite an annual use of around 45 million electrical detonators in the United States, there have been very few substantiated incidents of a commercial detonator firing accidentally because of RF-induced currents. Investigations tend to show that even the verified cases could have been avoided if the recommended safety procedures had been rigidly followed.

DO post warning signs within 1,000 feet of blasting sites stating that two-way radios must not be used because of blasting. NOTE: This is a federal regulation requirement.

DO make periodic checks of areas surrounding a fixed-blasting site for new sources of RF energy.

DO place signs throughout a blast site warning users of two-way radios that they must not be used while on-site. Support this with checks of incoming vehicles and the implementation of a positive means of preventing the use of vehicular two-way radios. You could, for example, physically disconnect the PTT switch/microphone, disconnect the set from the battery, or place a "stop" or a piece of tape over the PTT/transmit switch to prevent its use or to remind the operator of the dangers.

Minimum Safe Distances

DO maintain the minimum safe distance (MSD) from CB and other mobile transmitters as shown below:

SAFE DISTANCES FROM CB AND OTHER MOBILE RADIOS

Transmitter Power into the Antenna in Watts	Frequency in Megahertz and Primary Band User	Frequency in Megahertz and Primary Band User	Frequency in Megahertz and Primary Band User	Frequency in Megahertz and Primary Band User	Frequency in Megahertz and Primary Band User
	1.6-3.4 Industrial MSD IN FEET	28-29.7 Ham Radio MSD IN FEET	35-36 & 42-44 Public Use 50-54 Ham Radio MSD IN FEET	144-148 Ham Radio 150.8-161.6 Public Use MSD IN FEET	450-470 Public Use & Above 800 Vehicular Cell Phones MSD IN FEET
5	30	70	60	20	10
10	40	100	80	30	20

Transmitter Power into the Antenna in Watts	Frequency in Megahertz and Primary Band User	Frequency in Megahertz and Primary Band User	Frequency in Megahertz and Primary Band User	Frequency in Megahertz and Primary Band User	Frequency in Megahertz and Primary Band User
	1.6-3.4	28-29.7	35-36 & 42-44 Public Use 50-54	144-148 Ham Radio 150.8-161.6	450-470 Public Use & Above 800
	Industrial	Ham Radio	Ham Radio	Public Use	Vehicular Cell Phones
	MSD IN FEET	MSD IN FEET	MSD IN FEET	MSD IN FEET	MSD IN FEET
50	90	230	180	70	40
100	120	320	260	100	60
180	170	430	350	130	80
250	200	500	410	160	90
500	280	710	580	220	120
600	300	780	640	240	140
1,000	400	1,010	820	310	180
10,000	1,240	3,200	2,600	990	560

DO maintain the minimum distances from CB Class D transmitters, as shown below.

Safe Distances from Class D CB Transmitters

A. Double sideband transmitter, 4 watts maximum transmitter power

Hand-held: MSD = 5 feet
Vehicular: MSD = 65 feet

B. Single sideband transmitter, 12 watts PEP (Peak Envelope Power)

Hand-held: MSD = 20 feet
Vehicular: MSD = 110 feet

DO maintain the minimum safe distances from commercial AM (amplitude modulation) broadcast transmitters operating in the frequency range 0.535 to 1.605 MHz, as shown below:

SAFE DISTANCES FROM COMMERCIAL AM BROADCAST TRANSMITTERS FREQUENCY RANGE 0.535–1.605 MHZ

Transmitter Power into Antenna in Watts	Minimum Safe Distance in Feet
Up to 4,000	800
5,000	900
10,000	1,300
25,000	2,000
50,000	2,900
100,000	4,100
500,000	9,100

DO maintain the minimum safe distance from international broadcast and amateur radio transmitters operating at frequencies up to 50 MHz, as shown below:

SAFE DISTANCES FROM INTERNATIONAL BROADCAST AND HAM RADIO TRANSMITTERS—FREQUENCY RANGE UP TO 50 MHZ

Transmitter Power into Antenna in Watts	Minimum Safe Distance in Feet
100	800
500	1,700
1,000	2,500
5,000	5,500
50,000	55,000

The above table assumes use of the blasting circuit layout shown at the start of this section.

DO maintain the minimum safe distance from VHF television and FM broadcast transmitters, as shown below.

SAFE DISTANCES FROM VHF TV AND FM RADIO STATIONS

Effective Radiated Power in Watts	FM Radio Station MSD in Feet	TV Channels 2-6 MSD in Feet	TV Channels 7-13 MSD in Feet
Up to 1,000	800	1,000	600
10,000	1,400	1,800	1,000
100,000	2,600	3,200	1,900
316,000	3,400	4,300	2,500
1,000,000	4,600	5,800	3,300
10,000,000	8,100	10,200	5,900

DO maintain the minimum safe distance from UHF (ultra-high frequency) television transmitters, as shown below:

SAFE DISTANCES FROM UHF TV TRANSMITTERS

Effective Radiated Power in Watts	Minimum Safe Distance in Feet
Up to 10,000	600
1,000,000	2,000
5,000,000	3,000
100,000,000	6,000

DO maintain the minimum safe distance from maritime radio navigational radar transmitters, as shown below.

SAFE DISTANCES FROM MARITIME RADIO-NAVIGATIONAL RADARS

Type of Vessel	Typical ERP in Watts & Wavelength	MSD in Feet (except where stated)
Small Craft	500/3cm	20
Harbor Craft, Riverboats	5,000/3cm	50

Large Commercial Shipping	50,000/3cm and 10cm	300
"Rule of Thumb" Safety Distance if Uncertainty Exists as to the Wavelength and ERP of the Radar Beam	???	1,000
Large Commercial Shipping with Long-Range Radar	1,000,000/0.2m	Hazardous to within 1 Mile (Consult Local Authority if Blasting Location Indicates a Risk from Such Vessels).

DO maintain the minimum distance from radio-navigation beacons, as shown below:

SAFE DISTANCES FROM RADIO-NAVIGATION BEACONS

Beacon Type	Power in Watts	Frequency in MHz	MSD in Feet
Guide Slope	15	315	25
Localizer	100	110	110
Loran-C	1,000,000	0.1	650
Omega	10,000	0.01	45
VOR	100	110	110

DO select UHF frequencies (typically 450 MHz to 470 MHz) for communications equipment if possible. RF pickup by a typical electrical blasting circuit is far less efficient at these frequencies than at the lower frequencies.

DO avoid the use of large loops in electrical blasting circuits by running the wires next to each other or using twisted pairs.

DO keep any unavoidable loops small and oriented broadside onto known transmitting antennas.

DO keep lead lines out of the beam of directional RF systems such as microwave or radar arrays.

DO keep lead lines on the ground wherever possible.

DO contact the commander of any military base near which you are to blast to check on the location, direction, and power of any transmitters.

SPECIFIC BLASTING TECHNIQUES

Agricultural Blasting

Agricultural blasting (stumping, ditching, boulder blasting, land clearance, etc.), should *not* be attempted by inexperienced personnel. Note also that many states require that all users of explosive materials hold a valid blaster's permit, as well as a purchase and storage permit, even if the blasting is to be done on their own property.

Ditch Blasting (Ditching)

General

DO be familiar with the moisture content and other ground conditions in the area to be blasted.

DO use knowledge of the soil type and conditions to determine the amount and type of explosive material and the depth and spacing of holes to be used in a test blast.

DO note that although there is no real limit, as such, to the length of a ditch blast, in practice it has been determined that it is best to restrict the maximum length to around 300 to 500 feet per blast.

DO cut and clear timber and brush (including overhanging timber, which will otherwise catch debris and allow it to fall back into the ditch) from the area of the ditch line.

DO use a centerline of poles or rope to ensure the straight layout of blast holes.

DO load cartridges in vertical holes of the same depth.

DO ensure that all holes, and especially larger diameter holes, are stemmed. Introduce stemming material if natural stemming (water, mud) is not adequate.

Propagation Ditching

In this technique only one blast hole is primed. The concussive wave from this explosion passes through the ground to the next charge, causing it to detonate sympathetically and so on along the line.

DO use this technique when the ground is sufficiently wet and soft to permit it.

DO use a relatively sensitive explosive material for propagation ditching, e.g., ditching dynamite. Ditching dynamite is 50-percent straight nitroglycerine dynamite, usually supplied in the form of 1/2-pound0 8" x 1.5" cartridges.

Individual Primed Hole Ditching

This technique will be used when the ground is too hard and dry to permit propagation ditching. Here each hole contains a cartridge primed with either an electrical detonator, nonelectric detonator, or detonating cord.

Electric Detonators

DO connect electric detonators in series or series-in-parallel.

DO fire them simultaneously.

DON'T use more detonators than the blasting machine in use is capable of firing simultaneously.

Nonelectric Detonators and Shock Tube

DO connect the shock tube to the correct firing console or a trunkline of detonating cord.

DO fire the holes simultaneously.

Nonelectric Detonators and Gas Initiation

DO connect the holes in series or series-in-parallel.

DO fire the holes simultaneously with the correct gas initiation machine.

Detonating Cord

DO connect the holes to a single trunkline of detonating cord fired with an electric or nonelectric detonator.

LOADING TECHNIQUES

"Punch Bar" Loading Technique

DO use a single line of equal-spaced holes along the centerline of the proposed ditch.

DO ensure that the top of the cartridge (or the top of the last cartridge inserted if multiple cartridges are used) is not more than 12 inches below the surface in the case of dry ground or not more than 4 inches in the case of wet and soft ground.

DO place two extra holes to form a "Y" shape, with the last hole in a single-line ditch to create a square, rather than a round end.

Punch Bar Loads

The following table assumes normal conditions. To confirm these calculations, 25- to 50-foot-long test shots should be made. Local variations in soil type, etc., will often call for adjustment.

PUNCH BAR LOADS

Number of 1.25" x 8" Cartridges Per Hole	Depth to Column Top in Inches	Resulting Ditch Depth in Feet	Width of Ditch at Top in Feet	Distance Between Holes in Inches	Pounds Per 100' of Ditch
.5	6-8	1.5–2	4–5	12	25
1	6-12	2.5–3	6	15	40
2	6-12	3–3.5	8	18	67
3	6-12	4–4.5	10	21	86
4	6-12	5–5.5	13	24	100
5	6-12	6–6.5	16	24	125

"Posthole" Loading Technique

DO use this technique when it is desired to blast a ditch greater than 6 feet in depth.

DO use large (4-inch or 5-inch diameter) cartridges to speed up loading if the size of the ditch calls for a large quantity of explosive material.

DO place the charges in a single line at a depth equal to one-half or one-third of the required depth of the ditch. This technique will usually create a ditch with a bottom width equal to its depth and a top width some three times greater than the depth.

Posthole Loads

The following table assumes horizontal conditions. To confirm these calculations, 25- to 50-foot-long test shots should be made. Local variations in soil type, etc., will often call for adjustment.

POSTHOLE LOADS

Quantity of Explosive Material Per Hole in Pounds	3	5	10	15	25	50
Distance Between Holes in Feet	3	3.5	4	4.5	5	6

Resulting Ditch Depth & Bottom Width in Feet	4	5	6	6	8.5	12
Resulting Top Width in Feet	12	15	18	21	25.5	36
Depth of Load in Feet	2 2/3	3 1/3	4	4 2/3	5 2/3	8
Borehole (Posthole) Diameter in Inches	4	4	4	4	8	8
Weight of Explosive Material Per 100 Feet in Pounds	100	142.5	250	333	500	833
Material Moved Per 100 Feet in Cubic Yards	118	185	266	363	533	1,067

DO be aware that ditch blasting can produce large amounts of debris, which may be carried considerable distances by strong winds.

DO be aware that such material may present a risk to nearby dwellings, vehicles, personnel, etc.

DO ensure that all blasting personnel are upwind and behind or beneath protective cover.

Boulder Blasting

Blockholing

DO use the rule-of-thumb ratio of one-half to one-quarter of a 1.25" x 8" cartridge per cubic yard of rock to be broken.

DO drill the hole about halfway into the boulder before loading a primed charge of explosive material.

DO stem the charge adequately with damp sand or clay.

DO use a nonelectric detonator and safety-fuse firing system only when firing a single hole. If firing multiple charges, use electric, shock tube, detonating cord, or gas-initiation systems.

DO take adequate precautions to protect personnel from the effects of flyrock and spalling.

Mudcapping

DO use the rule-of-thumb ratio of 2 to 3 pounds of explosive material per cubic yard of rock to be broken.

DO place the primed charge in a natural depression on the boulder or in some other position where good contact is ensured.

DO cover the charge with at least 4 to 6 inches of wet mud to provide adequate confinement.

DO use a nonelectric detonator and safety fuse firing system only when firing a single hole. If firing multiple charges, use electric, shock tube, detonating cord or gas-initiation systems.

DO take adequate precautions to protect all personnel and equipment from the effects of airblast and flyrock and spalling.

POND/DRAINAGE/IRRIGATION BLASTING IN WET LOCATIONS

DO apply the rule-of-thumb ratio that 1 pound of explosive material will remove 1 cubic yard of earth.

DO use the punch-bar method of ditching for blasting shallow ponds of 2 to 3 feet in depth. Use parallel lines of holes spaced around 30 inches apart, each hole being loaded with a single 1.25" x 8" cartridge.

DO use the posthole technique if deep, large ponds are required.

DO place an extra hole, primed with two cartridges, between each primary row at the center of the pond to aid in scouring and to prevent debris from falling back in on itself.

DO use ditching dynamite if the propagation technique is to be used.

DO ensure that all explosive materials used are either water-resistant or adequately waterproofed. An ANFO (ammonium nitrate fuel oil) charge with installed HE (high-explosive) primer sealed in a plastic bag is commonly used.

DO take adequate precautions to protect personnel and equipment from the effects of flying debris. Personnel must be at a safe distance and under cover.

STUMP BLASTING

DO initiate stumping blastholes with detonating cord or electric detonators whenever possible.

DO apply the rule-of-thumb principle that a new "green" stump will require more explosive material than an old or rotting one.

DO determine the size of the stump by measuring its diameter 1 foot above the ground.

DO use the type of soil beneath the stump, as well as the size and type of stump itself, as an indicator of how much explosive material will be needed. Sandy or loose soil offers less confinement than clay or loamy soils.

DO bury the charge deeper beneath the stump if the soil is particularly light.

DO slit and tamp explosive charges (*not including the primer charge*) inserted beneath the stump so that they will compact and fill the cross section of the hole.

DO stem the hole(s) with clay or heavy loam-type soil, if available.

DO ensure that all personnel are at a safe distance and undercover before firing.

Stumping Cartridge Estimation

The following table assumes the use of standard 1.25" x 8" cartridges and firm, dense soil in areas of the United States other than the Pacific Northwest. Because of the widely variable nature of stumps and soil, this chart should be considered a rough guide only.

STUMPING CARTRIDGE ESTIMATION

Diameter of Stump Measured 1 Foot Above the Ground in Inches	Cartridges Needed Green Stump	Cartridges Needed Old/Established But Still Solid Stump	Cartridges Needed Partially Rotten Stump
6	2	1.5	1
12	4	3	2
18	6	4	3
24	8	6	4
30	10	7	5
36	13	9	6
42	16	12	8
48	20	15	10

Pacific Northwest Stumping Data

Stump diameters in this area are usually measured 4 feet above the ground. It has been found that green or fairly young hemlock stumps will need around one standard 1.25" x 8" cartridge for each inch of diameter; spruce will need around two-thirds of a cartridge; maple and alder half a cartridge. Cedar will need even less.

COAL MINING

Surface Mining

The type and nature of rock formations above the coal will influence considerably the type of explosive material used, its placement, and the diameter of blastholes. Typical blasthole diameters in this field range from

5 to 18 inches, with a common average being 6 to 9 inches. Blasthole depths will vary from 20 feet or less to 100 feet in the case of large operations.

ANFO is commonly used in coal-stripping projects, and in small operations will be available in thick, multi-walled paper sacks for pouring by hand. In large projects, the material will more usually be dispensed by a bulk transport truck. "Wet-hole ANFO" consists of water-resistant tubing containing a mixture of powdered ammonium nitrate, fuel oil, and nonexplosive materials and is used to "dry up" a wet blast hole or to raise the explosive material column above the waterline so that bulk ANFO can be added. Typical wet-hole cartridge sizes and weights are 5" x 25 pounds, 6" x 30 pounds, and 7" x 50 pounds, etc. Cartridged ANFO must have a specific gravity of at least 1.05 to ensure that it sinks properly in wet blast holes. However, as the ANFO mixture loses its sensitivity if the density exceeds 1.20, other ingredients are frequently added to maintain the correct balance. Such ingredients include coal dust and aluminum. Other water-resistant cartridges contain gels or slurries. Mixtures of emulsions, slurries, water gels, and ANFO (known as "heavy ANFO") may be encountered also. Where the water problem is minor and can be dealt with by pumping it away, a polyethylene liner (weighted at the bottom with dull cuttings) may be inserted to protect a subsequently loaded column of explosive material from the moisture.

Boreholes will usually be drilled to just above the cool seam (or actually to the coal seam) and backfilled so as to avoid smashing the coal. As the overburden may consist of alternating layers of soft and hard material, the charge types may be alternated accordingly or decking techniques employed. Decking techniques may also be employed in order to reduce the amount of explosive material fired in any one delay period, with a view toward lessening the effects of vibration on nearby residences. Vertical boreholes are commonly used for strip-mine blasting, but sometimes horizontal holes will be encountered, especially when the coal is thinly layered. Here a single row of holes will be drilled into the overburdern about 1 foot above the level of the coal seam. Such holes may be 100 feet deep. The back of the holes will be primed and the charges loaded. Extra boosters should be positioned every 10 feet. Bulk or cartridged explosives may be employed. A final booster or primer should be inserted at the end of the column before adding stemming. Stemming material is usually bagged sand or drill cuttings. Cast boosters are commonly used both as boosters proper in conjunction with columns of cartridged blasting agents and as primers with non-cap-sensitive material.

Even though a single primer will usually be capable of bringing a blasting agent up to a steady state velocity, common practice is to use an "insurance" primer, especially in deep holes. Certain regulations and company standard operating procedures insist on the use of two primers in each hole. In the context of boostering long cartridge columns in wet, deep holes, serious thought should be given to employing a detonating cord downline with a booster after every two cartridges. This technique ensures that a booster is in contact with each cartridge.

Underground Mining

Only Mine Safety and Health Administration (MSHA) approved, permissible explosives are allowed for blasting in underground coal mines, and circuit testing and blasting machines must be type-approved for underground coal mine use. The use of detonating cord, black powder, safety fuse, and electric detonators with aluminum shells or legwires is forbidden in such locations. The interval between successive delay periods must be at least 50 milliseconds. Only noncombustible stemming must be used.

Permissible explosives are available in nongelatinous or gelatinous form in various densities and detonation velocities. Permissible emulsions and gels may be encountered also. The commonest cartridge diameter is 1.25 inches, and lengths are usually 8 inches, 12 inches, 16 inches, or 2 feet.The technique most usually employed in such mining operations is to cut (with a machine) a shearing (vertical cut) or kerf (undercut) in the coal seam, thereby providing a relief for the boreholes. An alternative is to dislodge coal with explosives only. This is known as shooting "off the solid." In the case of the former, a 1 7/8-inch diameter auger is generally used to drill the boreholes which should not be nearer than 6 inches to the back of the shearing or kerf. Any holes that do extend beyond the back of the shearing or kerf must be backfed with noncombustible stemming material to a point 6 inches less than the depth of the cut. Holes must never extend laterally beyond the kerf as holes in the solid beyond the ribline could cause a shot to blow out.

Boreholes must have at least 2 feet of burden in all directions except where restricted by the coal seam itself. A primer, with the base end of the detonator pointing toward the collar of the hole, is to be inserted into the hole first. The rest of the cartridges will be file loaded and pushed fully home with a loading pole. Permissible explosives must never be slit or tamped.

The total weight of explosives loaded in a borehole in an underground coal mine must not exceed 3 pounds, and the total weight of explosives in a borehole less than 6 feet deep must be reduced by 1/2-pound for each foot of borehole less than 6 feet that the hole in question measures.

Holes 4 feet or deeper must be stemmed for at least 2 feet. Holes less than 4 feet in depth must be stemmed for at least half the depth of the hole.

Handling Misfires in Coal-Mining Operations

Wash the stemming and explosive material out of the hole with water and remove the explosives for subsequent destruction.

Or you can wash out the stemming with water, insert a new primer (ensuring that the base of the detonator is facing toward the charge), stem the hole again, and refire.

Or drill a new hole at least 2 feet away from and parallel to the misfire. Load, stem, and fire the new hole.

DO search the muckpile for unfired explosive material that may have been displaced from the misfire hole.

COYOTE BLASTING

This is a specialized technique employed when the terrain renders the usual technique of drilling unsafe, uneconomical, or impractical, usually by virtue of dangerously steep slopes and especially when the rock formations to be blasted in such locations are columnar or otherwise susceptible to displacement. Because of the amount of explosive used and the need to accurately determine their proper placement with regard to the precise nature of the rock and its particular formation, coyote blasting must never be attempted by inexperienced personnel.

The technique used is to cut small tunnels (known as

"coyote" drifts) into the rock and then to drive "wings" or "Ts" outward, perpendicular to the coyote drift and parallel to the rock face. Charges are designed and loaded into the wings in a manner calculated to ensure maximum, proper distribution of the explosive energy. Bagged or bulk ANFO charges primed with cast boosters or other cap-sensitive explosives are commonly used, initiation being via a double-detonating cord trunkline with numerous cross ties. The trunkline cords are run out of the main tunnel and protected with planks of wood, sandbags, or plastic tubing. The tunnel itself is then stemmed (backfilled).

MARINE OPERATIONS

DO store explosives and detonators in separate magazines on deck.

DO keep the magazines securely fastened to the deck and ensure that they are:

1. Separated from each other by a minimum of 25 feet.
2. At least 15 feet away from ducts exhausting hot air.
3. At least 10 feet away from unshielded radio equipment or antenna leads (see also section on RADHAZ).

DO remove only the explosive materials required to make up one charge at a time from the magazine.

DO ensure that the surface float of a positioned underwater charge is clearly visible to the shooter and assistant shooter at all times.

DO ensure that the float is at the prescribed distance from the operations vessel before the electrical detonator wires are connected to the firing line.

DO ensure that the charge is at the shooting point before

the assistant shooter unshorts his shorting plug.

DO ensure that at this point the assistant shooter communicates to the shooter, "Ready to fire."

DO ensure that the shooter unplugs his shorting plug in preparation for firing only when all the previous conditions are met.

AGENCIES ADMINISTERING REGULATIONS RELATED TO EXPLOSIVE MATERIALS

STATE AND LOCAL

Department of Mining and Minerals
Fire Marshal
Department of Transportation
Department of Labor and Industry
Sheriff

FEDERAL

Bureau of Alcohol, Tobacco, and Firearms (BATF)
Department of Transportation (DOT)
Mine Safety and Health Administration (MSHA)
Occupational Safety and Health Administration (OSHA)
Office of Surface Mining (OSM)
United States Coast Guard (USCG)

GLOSSARY OF COMMERCIAL EXPLOSIVES TERMS

AC—Alternating current.

Acceptor—A charge receiving an impulse from another exploding donor charge.

Activation Energy—The minimum amount of energy required to initiate a reaction in a chemical system.

Actuate—To cause a device to function as designed.

Adiabatic Heating—In thermodynamics an effect in which no heat enters or leaves the system. Adiabatic compression of a gas causes heating; adiabatic expansion causes cooling.

Adobe Charge—An unconfined or mud-covered charge fired in contact with a rock surface without use of a borehole.

Airblast—The acoustic transient or airborne shock wave generated by an explosion.

Ammonium Nitrate—An ammonium salt of nitric acid represented in the formula NH_4NO_3.

Amp(ere)—A unit of electrical current produced by one volt acting through a resistance of one ohm.

ANFO—Ammonium nitrate fuel oil; also known as "prills and oil." A blasting agent used widely in blasting operations. ANFO agents are characterized by low density and poor water resistance. Cartridged ANFO agents therefore incorporate crushed ammonium nitrate and densifying agents sealed in a water-resistant container. ANFO must not be mixed at underground locations. Premixed ANFO charges for use underground incorporate additives to lessen the dangers from fumes.

ANSI—American National Standards Institute. A non-government agency concerned with developing health and safety standards for industry.

Arming—The changing from a safe condition to a state of readiness for initiation.

Artificial Barricade—An artificial revetted wall or mound of earth with a minimum thickness of 3 feet.

Available Energy—The energy from an explosive material that is capable of performing useful work.

Backbreak—Rocks broken beyond the limit of the last row of holes in a blast. (Same as Overbreak.)

Ballistic Mortar—A laboratory test instrument used to measure the relative power of an explosive material.

Barricaded—The effective screening of a building or magazine containing explosives from another magazine or building by a natural or artificial barrier. A straight line from the top of any sidewall of the building containing the explosive materials to the eave line of any magazine or other building or to a point 12 feet above the center of a railway or highway shall pass through such a barrier.

Base Charge—The main explosive charge in the base of a detonator.

BATF—Bureau of Alcohol, Tobacco and Firearms.

Bench—A horizontal ledge from which boreholes are drilled vertically down into the material to be blasted. Benching is the process of excavating where a highwall is worked in steps or lifts.

Bench Height—The vertical distance from the top of a bench to the top of the next lower bench or to the floor.

Black Powder—An intimate mixture of sulfur, charcoal, and an alkali nitrate, typically sodium or potassium nitrate forming a deflagrating explosive compound.

Blasting—The firing of explosive materials for purposes such as moving material, breaking material, or generating seismic waves.

Blast Area—The area of a blast affected by flyrock, concussive effects, gasses, etc.

Blasthole—A hole drilled into the material to be blasted for the purposes of inserting explosive material. (Same as Drill Hole or Borehole.)

Blast Pattern—The pattern or plan of blastholes as laid out for blasting and thus an expression of the burden and spacing distances and their relationship to each other. (Same as Drill Pattern.)

Blast Site—The area in which explosive materials are handled during loading and including the perimeter of boreholes and 50 feet in all directions from loaded holes and empty adjacent holes that are to be loaded on the same shift. In underground mines, 15 feet of pillar or solid rib can be substituted for the 50-foot distance above ground.

Blaster—The qualified individual in charge of and responsible for the loading and firing of a blast. (Same as Shot Firer.)

Blasting Accessories—Nonexplosive devices and materials used in connection with blasting. Examples are cap crimpers, cartridge punches, tamping bags, blasting machines, etc.

Blasting Agent(s)—An explosive material meeting certain prescribed conditions regarding insensitivity to initiation. In the context of storage, Title 27, Code of Federal Regulations, Section 55.11 defines a blasting agent as any material or mixture consisting of fuel and oxidizer intended for blasting, not otherwise defined as an explosive. Provided that the finished product, as mixed for use or shipment, cannot be detonated by means of a No. 8 test blasting cap when unconfined. For transportation, Title 49 Code of Federal Regulations defines a blasting agent as a material designed for blasting which has been tested in accordance with Section 173.115a and found to be so insensitive that there is very little probability of accidental initiation to explosion or transition from deflagration to detonation.

Blasting Cap—A detonator designed to be initiated by safety fuse. (Same as Fuse Cap.)

Blasting Crew—A team of people assisting the blaster.

Blasting Galvanometer—A special electrical instrument used for testing for electrical continuity in electric detonators and firing circuits. Other instruments that may be used are blasting ohmmeters and blasting multimeters. If the equipment doesn't actually have the word blasting or blasters on it, it must not be used.

Blasting Log—A written record of a specific blast, often required by law.

Blasting Machine—Also known as an exploder. An electrical or electromechanical device providing electrical energy for the purpose of firing detonators in an electrical blasting circuit. The term is also used to describe certain equipment used in connection with nonelectric blasting systems. Hand-operated generator types and electronic capacitor-discharge (CD) types exist.

Blasting Machine Rheostat—A variable electrical resistance device used to simulate electrical detonator resistances for the purpose of testing generator-type blasting machines.

Blasting Mat—A covering of woven steel wire, rope, scrap tires, or other material used to cover blastholes for the purposes of preventing flying debris.

Blasting Vibrations—Energy from a blast manifesting itself in the form of vibrations that are transmitted through the earth away from the blast area.

Blend—A mixture of 1) a water-based explosive material matrix and ammonium nitrate or ANFO, and 2) a water-based oxidizer matrix and ammonium nitrate or ANFO.

Blockholing—The technique of breaking large boulders and rocks by loading and firing small explosive charges in holes that have been drilled into them.

Booster—An explosive charge, typically one of a high detonation velocity and pressure used in the explosive initiation sequence between the initiator or primer and a main charge.

Bootleg—That part of a drilled borehole that remains when an explosion does not break the rock completely to the bottom of the hole. (Same as Socket.)

Borehole—(Same as Blasthole or Drill Hole.)

Breakage—A word used to describe the size distribution of rock fragments created by a blast.

Bridgewire—A resistance wire that connects the legwires of an electrical detonator inside the shell. The wire is embedded in the detonator ignition charge.

Brisance—The shattering capability of an explosive material.

Bulk Mix—Explosive material prepared for use in bulk form without packaging and usually delivered by a purpose-made vehicle.

BMDE—Bulk mix delivery equipment. A vehicle with or without a special delivery device used for transporting explosive materials in bulk form for mixing on-site or loading directly into blastholes.

Bulk Strength—The strength per unit volume of an explosive material calculated from its weight strength and density.

Bulldoze—(Same as Adobe Charge, Mudcapping, and Plaster.)

Bullet-Resistant—In the case of magazine walls or doors, this is defined as being resistant to the penetration of a bullet of 150-grain M2 ball ammunition having a nominal muzzle velocity of 2,700 fps when fired from a .30-caliber rifle from a distance of 100 feet perpendicular to the wall or door. For magazine roofs or ceilings, the requirement is that they be constructed from materials comparable to those used in the side walls or other materials that are capable of resisting penetration of the bullet described above when fired at an angle of 45 degrees from the perpendicular. Tests to confirm bullet resistance will be conducted on test panels or empty magazines that will be required to resist 5 out of 5 shots placed independently of each other in an area at least 3 feet by 3 feet.

Bullet-Sensitive Explosive Material—Explosive materials that can be detonated by 150-grain M2 ball ammunition having a nominal muzzle velocity of 2,700 fps when the bullet is fired from a .30-caliber rifle at a distance of 1,000 feet and the material—at a temperature of 70° to 75°F—is positioned against a backing plate of 1/2-inch steel.

Bureau of Explosives—A bureau of the Association of American Railroads which the Department of Transportation may consult for recommendations on classification of explosive materials for the purpose of interstate transportation.

Burden—1) The distance from the blasthole to the nearest free face or the distance between blastholes measured perpendicular to the spacing, and 2) the total amount of material to be blasted by a particular hole, usually expressed in cubic tons or yards.

Bureau of Mines—The U.S. Bureau of Mines (USBM). A bureau of the Department of Interior active in promoting safety in coal mines.

Circuit—A completed path along which electrical current can flow. The term may also be encountered in the context of certain nonelectric blasting circuits.

Class A Explosives—An old U.S. Department of Transportation (DOT) definition meaning explosive materials that present detonating or other maximum hazard risks. Examples are blasting caps, dynamite, and nitroglycerin. The new classification is DIVISION 1.1 or 1.2.

Class B Explosives—An old DOT definition that presents a flammable hazard. Examples are propellant explosives, certain fireworks, and photographic flash powders. The new classification is DIVISION 1.2 or 1.3.

Class C Explosives—An old DOT definition referring to explosives that contain Class A or Class B explosives or both as a constituent part but in restricted quantities. The new classification is DIVISION 1.4.

Collar—The mouth or opening of a blasthole or shaft.

Column Charge—A charge of explosive material in a blasthole in the form of a continuous column.

Column Depth—The length of each portion of a blasthole filled with explosive materials.

Combustion—Exothermic oxidation reaction (the combination of substances with oxygen usually accompanied by flame, smoke, or sparks) with the oxygen being supplied by the atmosphere or from within the material itself.

Commercial Explosives—Explosive materials designed, produced, and used for commercial or industrial applications as opposed to military applications.

Confined Detonation—The detonating velocity of explosive material in a blasthole or a substantial container.

Connecting Wire—Wire used to lengthen the firing line or legwires in an electrical blasting circuit.

Continuity Check—A check made with test instruments (or visually) to confirm that an initiation system is continuous and unbroken and containing no breaks or faults that could cause failure or interruption of the initiation process.

Contour Blasting—A technique used to create smooth walls and reduce overbreak in underground blasting. Cushion holes are used that contain light, well-distributed charges, fired on the last delay period on the ground.

Core Load—The weight (expressed in grains of explosive per foot length) of the explosive cord in detonating cord.

Coupling—The extent to which explosive material fills the cross-section of a blasthole. Bulk-loaded charges would be completely coupled, untamped charges uncoupled.

Coyote Shooting—A special blasting technique wherein several relatively large, concentrated charges are fired in one or more small tunnels driven into a rock formation.

Crimp—1) The circumferential depression in a blasting cap shell that secures a sealing plug or sleeve. 2) The circumferential depression at the open end of a fuse cap or ignitor cord connector that secures the fuse. 3) The folded end(s) of paper covered explosive cartridges.

Crimper(s)—(See Cap Crimpers)

Crimping—The process of securing a fuse cap or ignitor cord connector to a section of safety fuse by compressing the metal shell of the cap against the fuse with a special tool known as a crimper.

Critical Diameter—The minimum diameter required to ensure propagation of a detonating wave at a stable velocity. The critical diameter is affected by confinement conditions, temperature, and pressure on the explosive material.

Current Leak—Any portion of the firing current bypassing any part of the blasting circuit via unintended paths.

Current Limiting Device—An electrical or electromechanical device used to 1) limit current amplitude, 2) limit the duration of current flow, 3) limit the total energy of the current delivered to an electric blasting circuit.

Cushion Blasting—A blasting technique employing cushion holes with reduced spacing that contain decoupled charges fired after the main charge. The object is to obtain smooth walls or slopes.

Cutoff—A break in the initiation or detonation path caused by external influences such as shifting ground or flyrock.

Date-Shift Code—A code appearing on the outside shipping containers and sometimes on the inner containers and wrappers of explosive materials to enable their identification. The code is required under BATF regulations.

DC—Direct current.

Decibel—A unit of air overpressure used in the measurement of airblast.

Decking—A technique of loading blastholes in which charges (known as decks or deck charges) in the same hole are separated by stemming or an air cushion.

Deck—An explosive charge separated from another in the same blasthole by stemming or an air cushion.

Decoupling—A technique of loading in which cartridged explosive charges much smaller than the diameter of the blasthole are used. The technique minimizes stresses exerted on the blasthole walls.

Deflagration—An ultrafast burning phenomenon.

Delay—1) A distinct pause of predetermined time between initiation or detonation impulses used to enable the separate firing of explosive charges. 2) Any device designed to produce a time lag between initial actuation and, for example, the arming of a device or detonation of an explosive charge.

Delay Blasting—The technique of initiating individual blastholes, rows of blastholes, or decks at predetermined intervals by using delay detonators or other delaying means. The opposite of instantaneous blasting.

Delay Detonator—A detonator (electric or nonelectric) designed so as to provide a predetermined lapse of time between the application of the firing impulse and detonation of the base charge.

Delay Element—The device in the delay detonator providing the predetermined time lapse.

Delay Interval—1) The elapsed time between the detonations of delay detonators in a delay series. 2) The elapsed time between successive detonations in a blast.

Delay Period—A designation given to delay detonators showing the relative or absolute delay time in a delay series.

Delay Series Detonators—A series (or range) of delay detonators for use in specific delayed firing operations. Two types, MS (millisecond, sometimes called SP or short-period) and LP (long-period) exist. The former have delays in the order of milliseconds, the latter in seconds.

Delay Tag—The tag, band, or marker on a delay detonator indicating the delay series to which it belongs, the delay period and/or the delay time of the detonator.

Delay Time—The time elapsed between the application of the firing stimulus and the detonation of the base charge in a delay detonator.

Density—The mass of explosive material per unit of volume, usually expressed in grams per cubic centimeter or pounds per cubic foot.

DOT—Department of Transportation. A cabinet-level agency of the U.S. federal government having responsibility for the regulation of transportation safety procedures relating to explosive materials and other hazardous material.

Detonating Cord—Flexible, waterproofed cord having a center core of high explosive material. Commercial types have core loads of 4.5 to 60 grains per foot length. Usually initiated via a detonator, detonating cord with a coreload in excess of 18 grains per foot can be initiated via knotted connections.

Detonating Cord Downline—The length of detonating cord extending within the blasthole from the surface down to an explosive charge.

Detonating Cord MS Connectors—Nonelectric delay devices (millisecond) used to delay blasts that are initiated by detonating cord.

Detonating Cord Trunkline—That line of detonating cord used to connect and initiate other lines of detonating cord.

Detonating Primer—A term used for transportation purposes meaning a device comprising a detonator and additional charge of explosive material assembled as a single unit.

Detonation—An ultrafast explosive decomposition wherein a heat-liberating chemical reaction maintains a shock front in the explosive material.

Detonation Pressure—The pressure created in the reaction zone of a detonating explosive.

Detonation Velocity—The speed at which a detonation wave moves through an explosive.

Detonator—A device containing a primary or initiating explosive used for initiating detonation in another explosive material. A detonator may not contain more than 10 grams of total explosive material by weight, excluding any ignition or delay charges. The term includes electric blasting caps (instantaneous and delay types), blasting caps for use with safety fuse, detonating cord delay connectors, and nonelectric instantaneous and delay blasting caps that use detonating cord, shock tube, or any other replacement for electric legwires. Unless specifically classified otherwise, detonators are Class A (Division 1.1 or 1.2) explosives.

Detonators Class C—Initiating devices that will not mass detonate when packed as for shipment.

Diameter—The cross-sectional width of a circular form.

Ditch Blasting—The creation of a ditch via the detonation of a series of explosive charges.

Ditching Dynamite—A sensitive, nitroglycerine-type explosive especially designed to propagate sympathetically from hole to hole in, for example, ditch blasting.

Donor—An exploding charge creating an impulse that, by impinging upon a second "acceptor" charge, causes it to detonate.

Dope—The individual, dry, nonexplosive materials that constitute a part of an explosive formulation.

Downline—The line of detonating cord or tubing in a

blasthole that will transmit the detonation wave from the trunkline or surface delay down to the primer.

Drill Hole—(Same as Blasthole or Borehole.)

Drilling Pattern—The location of a group of blastholes in relation to each other and the free face.

Dummy—A cylindrical lump of clay, sand, or other inert material used to confine or separate explosive charges in a blasthole.

Dynamite—Standard, ammonia, and gelatin (gelignite) dynamites exist. Granular (commonly green or brown and rubberlike in appearance) and solid types are available. Typically available in .5-pound, 8" x 1.25"-diameter cartridges. Military dynamite doesn't contain nitroglycerine and is therefore far safer to handle and store than commercial types. Old commercial dynamites will sweat nitroglycerine, causing oily stains on wrappers and storage cases. In this condition, it is extremely dangerous.

EBC—Electric Blasting Cap. Electric detonator.

Exploding Bridgewire. A detonator design wherein a short-duration, high-tension pulse causes the bridgewire to vaporize, thereby creating a shock wave that will detonate directly a surrounding charge of secondary HE. No primary explosive is therefore needed in the detonator.

EED—Electro-Explosive Device.

Electric Blasting Circuit—A blasting circuit designed for electrical initiation containing electrical detonators.

Electric Detonator—A detonator designed for initiation by electricity.

Electrical Storm—An atmospheric disturbance involving intense electrical activity producing lightning and powerful electromagnetic fields.

Emergency Procedure Card—Instructions to be carried in a vehicle conveying explosive materials that details what actions are to be taken in the event of an emergency arising.

Emulsion—An explosive material containing considerable oxidizer dissolved in water surrounded by an immiscible fuel, or droplets of an immiscible fuel surrounded by water containing considerable amounts of oxidizer.

Energy—A measure or indication of the potential for an explosive material to do work.

EOD—Explosive Ordnance Disposal.

ERP—Effective Radiated Power.

Exothermic—Characterized by or formed with evolution of heat. An exothermic reaction is a chemical reaction in which heat is liberated, the products being known as exothermic compounds.

Exploder—Another term for a blasting machine, usually referring to a button- or switch-operated battery and capacitor-based device used to fire electrical detonators.

Explosion—A chemical reaction characterized by the extremely rapid expansion of gases.

Explosive—Any chemical mixture, compound, or device, the purpose of which is to function by explosion.

Explosiveness—The speed or the extent to which an explosive releases its energy when subjected to a specific stimulus.

Explosive-Actuated Device—A tool or mechanical device put into operation by explosives. Examples are explosive cable cutters, jet tappers, and jet perforators.

Explosive Charge—The amount of explosive material used in a given device, blasthole, coyote tunnel, etc.

Explosive Loading—The amount of explosive used per unit of rock. Also known as the powder factor.

Explosive Materials—A term referring to all types of explosives, detonators, ignitors, detonating cord, safety fuse, squibs, etc. A list of materials falling under this heading is issued at least annually by the director of BATF as part of Title 18 USC Chapter 40, *Importation, Manufacture, Distribution, and Storage of Explosive Materials*. DOT classifications of explosive materials of the type used in commercial blasting projects are not identical to the statutory definitions of the Organized Crime Control Act of 1970, Title 18 USC, Section 841. To achieve consistency in transportation, the definitions of DOT in Title 49 Code of Federal Regulations Parts 1-999 subdivides the materials into four classes*:

1) *Class A Explosive* (now redesignated as Division 1.1 or 1.2 materials) detonating or otherwise maximum hazard materials.
2) *Class B Explosives* (now redesignated Division 1.2 or 1.3 materials) those posing a flammable hazard.
3) *Class C Explosives* (now redesignated Division 1.4 materials) those presenting a minimum hazard.
4) *Blasting Agents* (now redesignated Division 1.5 materials) materials designed for blasting that have been tested in accordance with Section 173.115a and found to be so insensitive that there is very little probability of accidental initiation to explosion or transition from deflagration to detonation.
* Division 1.6 materials have no other applicable class.

Explosive Oil—Liquid explosive sensitizers such as nitroglycerine, EGDN (ethylene glycol dinitrate), and metriol trinitrate.

Explosive Power—The work capacity of a high explosive calculated on the basis of the heat and gas generated compared to that of a standard explosive or by tests such as the lead block test.

Extra Dynamite—A dynamite in which a packet of nitroglycerine is replaced by ammonium nitrate in sufficient quantity to result in the same weight strength. Also known as Ammonia Dynamite.

Extraneous Electricity—Any electrical energy other than that desired for testing or initiating purposes present at a blast site and liable to enter the electrical blasting circuit. The term includes RF energy and static electricity.

Fertilizer-Grade Ammonium Nitrate—A specific grade of ammonium nitrate as defined by the Fertilizer Institute.

Fire Extinguisher Rating—A National Fire Code rating appearing on extinguishers in the form of a number indicating the extinguisher's relative effectiveness and a letter indicating the class of fire for which it has been proven effective.

Fire-Resistant—Generally, any construction designed to offer reasonable protection against fire. In the case of exterior magazine walls made of wood, this means a resistance equivalent to that provided by sheet metal of not less than 26 gauge.

Fireworks—Pyrotechnic articles designed for the purpose of producing entertaining audiovisual effects.

Firing Current—An electrical current of sufficient magnitude and duration to energize a given electrical blasting circuit.

Firing Line—The wires connected to the electrical power source in an electrical blasting circuit.

First Fire—An incendiary/pyrotechnic composition used to ignite a subsequent or the main charge.

Flammability—The relative ease with which an explosive material can be ignited by flame and heat.

Flare—A pyrotechnic device designed to produce a single source of bright light.

Flashover—The sympathetic detonation between charges in a blasthole or blastholes in close proximity to each other.

Flashpoint—The lowest temperature at which vapors from a volatile combustible material ignite in air when exposed to flame. It is determined by testing in special equipment.

Flyrock—Rock and debris propelled from a blast by the force of the explosion.

Forbidden Explosives—Explosive materials which are prohibited from transportation by common, contract, or private carriers by rail freight, rail express, highway, air, or water in accordance with DOT regulations.

Fragmentation—The breaking of solid mass into small pieces by blasting.

Free Face—A rock surface exposed to air or water that provides room for expansion when fragmented. Also known as the Open Face.

Fuel—A substance that can react with oxygen to facilitate combustion.

Fumes—Gaseous products from, in this context, an explosion. Explosive materials are given a fume classification based on the amount and type of poisonous or toxic gases they can produce.

Fuse Cap—A detonator designed to be initiated with safety fuse. Also called a Blasting Cap or simply Detonator.

Fuse Cutter—A mechanical tool for cutting safety fuse smoothly and cleanly at right angles to its axis.

Fuse Lighters—Pyrotechnic devices used to rapidly and reliably light safety fuse.

Gap Sensitivity—The maximum distance across which a detonation wave will jump and initiate another charge.

Gas Initiation—An initiation system in which detonators are initiated by an explosive gas conveyed to the detonators via thin, plastic tubing. Each detonator has two empty plastic tubes crimped into its open end. Detonators may be interconnected in series or series-in-parallel circuits with lengths of special plastic tubing and nonexplosive connectors. The initiation wave travels at around 8,000 fps. The entire system can be checked for leaks or blockages prior to firing. Special charging and firing units are employed.

Gelatin Dynamite—A very water-resistant type of dynamite, characterized by its gelatinous or plastic consistency.

Geology—Referring to the nature and configuration of

rock in, for example, rock formations, that will have an influence on blast designs.

Grain—1) A measurement of weight being 0.0648 grams in the troy and avoirdupois systems. In the avoirdupois system 7,000 grains are equal to 1 pound (0.45 kgs).
2) In the context of propellants, a piece of the propellant material itself, often of a specific geometric shape and regardless of its size. Thus, one "stick" of solid propellant of the type used in an antitank rocket would be termed a grain.

Ground Fault—An electrical path between any part of an electrical blasting circuit and earth (ground).

Ground Vibration—A shaking of the ground caused by shock waves moving outward from a blast. Measured in inches per second of particle velocity.

Hangfire—The detonation of an explosive charge or charges at some unpredicted time after the planned and intended firing time.

Hardwood—Red or white oak, hard maple, ash, or hickory free from all defects.

HE—High explosive.

Hertz—A measurement of frequency meaning cycles per second. Abbreviated as Hz (single cycles per second). Also KHz (thousands of cycles per second), and MHz (millions of cycles per second).

High Explosives—Explosives characterized by a very high reaction rate, the presence of a detonation wave in the explosive material, and the development of extremely high pressures.

Highwall—An almost vertical face of the edge of a ledge, bluff, or bench, or a surface excavation.

Hole Diameter—The cross-sectional width of the blasthole.

IED—Improvised Explosive Device.

Igniting Agent—A spark, flame, hot solid, etc.

Ignitor Cord—A pyrotechnic cord of small diameter burning at constant rate with an external flame. Used to light a series of safety fuses. Used with ignitor cord connectors, which are small metal capsules containing an ignition compound. The connectors are crimped to the safety fuse and the ignitor cord is inserted under the lip of the connector and the lip pressed closed. As the ignitor burns down, it ignites each connector in turn, thereby lighting the lengths of fuse.

IME Fume Classification—A classification indicating the amount of toxic and poisonous gas produced by an explosive or blasting agent. There are 3 classes:

1) Less than 0.16 cubic feet of poisonous gas per 1.25" x 8" cartridge
2) 0.16 to 0.33 cubic feet
3) 0.33 to 0.67 cubic feet.

IME—Institute of Makers of Explosives. The safety organization of the commercial explosives industry in the United States and Canada. It is a nonprofit organization primarily concerned with the safety and protection of employees, users, the public, and the environment in relation to the manufacture, storage, transportation, handling, and use of explosive materials.

Incendivity—The property of an igniting agent indicating that it is of sufficient intensity to ignite flammable material or explosive gases.

Inhabited (Building)—A building regularly occupied in part or in whole as a habitation for humans, also any church, school, train station, store, or other structure where people are accustomed to assemble, not including any building or structure occupied in connection with the manufacture, storage, transportation, or use of explosive materials.

Initiation—The commencement of detonation or deflagration in an explosive material.

Initiator—Any detonator, detonating cord, or similar device used to initiate detonation or deflagration in an explosive material.

Instantaneous Detonator—The opposite to a delay series detonator, i.e., a detonator which fires effectively instantly upon receiving the firing stimulus.

Inventory—A listing of all explosive materials stored in a magazine.

Issuing Authority—The specific government agency, body, office, or official endowed with the authority to issue permits or licenses.

Kelly Bar—A hollow bar affixed to the top of the drill column in rotary drilling. Also known as a Grief Joint, Kelly Joint, and Kelly Stem.

Lead Wires/Lines—(Same as Firing Line.)

Leakage Resistance—The electrical resistance between the ground and an electrical blasting circuit.

Leg Wires (Legwires)—The two wires (two single or one duplex) leading from an electrical detonator.

Liquid Fuels—Any fuels in a liquid state. Often mixed with oxidizers to create explosive materials.

Loading—1) Placing explosive material in a blasthole or against the material to be blasted. 2) The explosive material itself when so installed or positioned.

Loading Density—The weight of explosive material loaded per unit length of blasthole occupied by that explosive material. Expressed in pounds/foots or kilograms/meters of blasthole.

Loading Pole—A nonmetallic pole used to aid in the placing, moving, and tamping of explosive charges in a blasthole.

Low Explosives—Explosives characterized by a slow rate of reaction, deflagration, and the development of low pressures.

Magazine—Any structure or container approved for the storage or explosive materials, not including any structure used for the manufacture of explosives.

Magazine Keeper—The person who is in charge of the magazine and responsible for the safe and proper storage of explosive materials, inventory, and the immediate area.

Main Explosive Charge—The explosive charge performing the bulk of the work in a blast.

Manufacturing Codes—Codified markings on explosive materials packing indicating the date of manufacture and other information.

Mass Detonate—An effect wherein a single unit, or any part of a larger amount, of explosive material explodes and thereby causes the remainder of the units or the rest (or a substantial part of the rest) of the explosive material to detonate simultaneously. In the context of detonators, the definition of mass detonate means that 90 percent of the units, or more than 25 grams of the explosive material in the shipping container, explodes effectively simultaneously.

Maximum Recommended Firing Current—The highest electrical current that can be used to safely and reliably initiate an electrical detonation.

Millisecond—A measurement of time being one-thousandth of a second.

Miniaturized Detonating Cord—Detonating cord having a core load of around 2.4 grains per foot length. It is used with special instantaneous starters, surface delays, and in-hole delay detonators. It is not compatible with nitroglycerine-sensitized explosives.

Minimum Recommended Firing Current—The smallest electrical current that can be used to safely and reliably initiate an electrical detonation.

Misfire—1) A specific blasthole of a blast that fails to detonate as planned. 2) The explosive material itself that failed to detonate.

MS Connectors— Millisecond connectors. (Same as Detonating Cord MS Connectors.) Nonelectric devices used for delaying blasts initiated by detonating cord. Delay times are in the milliseconds range.

MSHA—Mine Safety and Health Administration. A Department of Labor agency concerned with publicizing and enforcing health and safety regulations in the area of mining.

MSHA Approval—An official document stating that an explosive or explosive unit has met MSHA requirements and authorizing an approval marking that indicates the explosive or unit is "permissible."

Muckpile—The heap or pile of broken material resulting from a blast.

Mudcapping—(Same as Adobe Charge, Bulldozing, and Plaster.)

Multiple Path Trunkline System—Duplication or repetition of trunkline sections providing alternate paths of initiation.

Munroe Effect—A localized concentration of shock wave energy at the target achieved by specific charge shape or distance of the charge at point of detonation from the target.

NFPA Standards—National Fire Protection Association Standards. Standards for explosive materials and ammonium nitrate issued by the NFPA.

NSC—National Safety Council. A nonprofit organization in the United States charged by Congress to report regularly on the causes of accidents and ways in which they may be prevented.

Natural Barricade—In the context of safe storage distances, these are any natural ground features, such as hills or timber of sufficient density, that obscure the surrounding exposures requiring protection.

Nitrate—1) Salt of nitric acid, or 2) to treat, impregnate, or cause to interact with nitric acid.

Nitroglycerine—Thick, clear-to-yellow/brown liquid that is extremely powerful and shock-sensitive. Freezes at 56°F. Far less shock-sensitive when frozen.

Nonelectric Detonator—A detonator not requiring an electrical current to function.

Nonsparkirig Metal—A metal that when struck against

rock, hard surfaces, or other metals will not spark. Used in the manufacture of crimpers and loading poles.

Overbreak—(Same as Backbreak.)

Overburden—The material on top of a deposit to be blasted or mined.

Oxidizer—A substance (for example, a nitrate) that will readily give up oxygen or other oxidizing substances to encourage the combustion of organic matter or other fuel.

Oxygen Balance—The percentage of oxygen in an explosive material or an ingredient of that explosive material in excess of (or less than) that amount needed to produce ideal reaction products.

Parallel Blasting Circuit—A multiple detonator electrical blasting circuit in which one legwire of each detonator is connected to one of the wires from the firing current source and the other legwire is connected to the other wire from the firing current source. The term may be encountered in the context of certain nonelectric blasting circuits.

Particle Velocity—In this context, a measure of the speed of motion of ground particles when excited by wave energy from an explosion.

Parting—1) A joint or crack in a rock formation. 2) A mass of rock located between two seams of coal, for example.

PBX—Plastic Bonded Explosive. Explosive material mixed with a plastic bonding agent and thus forming both the body (and often other parts) and the explosive charge in a given explosive device (typically of a mine).

Pellet Powder—Black powder in the form of compressed cylindrical pellets 2 inches long x 1.25 inch in diameter.

Permissible Diameter—The smallest diameter of a particular permissible explosive that may be used.

Permissible Explosives—Explosive materials approved for use in gassy and dusty atmospheres typically found underground.

PETN Explosive—Pentaerythritol Tetranitrate. A secondary high explosive commonly used in detonating cord.

Placard—A sign carried on the outside of vehicles transporting explosive and other hazardous materials to indicate the nature of the load and, frequently, the actions that should be taken in the event of an emergency.

Plaster—(Same as Mudcap and Mudcapping.)

Pneumatic Loading—Loading blastholes with explosive materials, compressed air being used as the loading or conveying force.

Posthole Loading—A blasting technique used to create deep ditches, for example. Larger than usual blastholes are required, and so posthole diggers are frequently used, hence the term.

Powder—A synonym for black powder or other explosive materials.

Powder Punch—(Same as Cartridge Punch.)

Powder Factor—(Same as Explosive Loading Factor.)

Power Source—The source of electrical current in electrical blasting circuits.

Preblast Survey—A documented examination of the preblast condition of structures near an area in which a blast is to occur.

Premature Firing—The unplanned detonation of an explosive charge before the intended firing time.

Presplitting—A blasting technique used to obtain smooth contours by firing a single row of blastholes prior to the rest of the holes in a given blast pattern.

Prilled Ammonium Nitrate—Ammonium nitrate in pelleted form.

Primary Blast—A blast used to move and fragment material with a view to subsequent blasting.

Primary Explosive—Explosive easily initiated by a small stimulus.

Primer—1) Any cartridge, unit, or package of explosive material containing either a detonator or a length of detonating cord to which a detonator is attached for the purpose of initiating the detonating cord, the whole assembly used to initiate other explosives or blasting agents. 2) A small, usually metal housing filled with impact- or stab-sensitive and flame-producing chemicals used to ignite the propellant in, for example, small arms cartridges. (Also called a cap.)

Propagation—The detonation of an explosive charge via an impulse from another attached, adjacent, or nearby explosive charge.

Propellant Explosive—An explosive material functioning normally by deflagration and being used for propulsion purposes. The material's susceptibility to detonation will determine its status as either a Class A (Division 1.2 or 1.2) or Class B (Division 1.2 or 1.3) explosive.

Propellant-Actuated Power Device—A tool or mechanical device actuated by a propellant or releasing and directing work via a propellant charge.

Pyrotechnics—Combustible or explosive articles and compositions producing visual or audible effects manufactured primarily for the purposes of entertainment.

Quantity-Distance Table—A table detailing the minimum recommended distances between stores of explosive materials of various weights to other specific locations.

RADHAZ—Acronym for Radio Frequency Hazard.

Radio Frequency (RF) Energy—Electromagnetic wave energy occurring in the radio frequency spectrum.

RF Transmitter—A transmitting system producing radio frequency energy.

Receptor—A charge of explosive material receiving or intended to receive an impulse from another exploding charge.

Relief—The distance from a blasthole to the nearest free face.

Resistance—A measure of the opposition to the flow of an electrical current expressed in ohms and abbreviated as Ω.

Rotational Firing—A blasting technique involving predetermined delays in such a manner that the charges successively move the burden into the void created by the previous explosion.

Round—A group of blastholes intended to be fired in a continuous series.

Safety Fuse—Flexible cord containing flammable material used to convey flame at a predictable, uniform rate from the point of ignition to the cut end, usually for the purpose of initiating a fuse cap attached to that end. Safety fuse may be used without a fuse cap to initiate certain deflagrating explosive material. Safety fuse manufactured in the United States has a burning rate of around 40 seconds per foot. Because of differences in manufacturing tolerances, variations in storage conditions, and the possible effects of weather, etc., an initial test burn on a 3-foot length from the roll to be used will always be made before safety fuse is used to determine its precise rate of burn. This should be supported by frequent check-test burns during the blasting operation. A minimum of 3 feet should be used with each detonator.

Scaled Distance—A factor that associates similar blast effects from varying weight charges of explosive material at various distances. Scaled distances for blasting effects are obtained by dividing the distance in question by a fractional power of the weight of the explosive material.

Secondary Blasting—Blasting boulders produced by a primary blast with a view to decreasing their size.

Secondary Explosive—An explosive that can be detonated by a substantial stimulus (i.e., when struck by a detonating wave) but not when ignited or heated.

Semiconductive Hose—Hose used for the pneumatic loading of explosive materials having an electrical resistance high enough to limit the flow of stray electrical currents to safe levels but not so high as to prevent leakage of static electricity to ground (earth). Technically, it is those having a resistance of not more than 2 million ohms (2 megohms) over its total length and not less than 1,000 ohms per foot length.

Sensitiveness—A measure of an explosive material's cartridge to cartridge propagating ability under specific test conditions. Expressed as the distance through air at which a primed half-cartridge will detonate an unprimed half-cartridge.

Sensitivity—A physical characteristic of explosive material being an indication of the ease (relative to a standard) with which it can be initiated or ignited with a particular external stimulus.

Separation Distances—The minimum recommended distances between accumulations of explosive materials and other specific locations.

Sequential Blasting Machine—(Same as Sequential Timer.) A blasting machine (exploder) designed to fire separate series of detonators at accurately predetermined intervals.

Series Blasting Circuit—A multiple-detonator electrical blasting circuit in which there is one continuous route for the firing current for all the caps in the circuit.

Series in Parallel Blasting Circuit—A multiple detonator electrical blasting circuit in which the detonators are divided into two or more balanced groups that are connected in series and in parallel.

Shaped Charge—Any type of explosive charge formed or cast into a specific shape to obtain a localized concentration of shock-wave energy.

Sheathed Charge—A device comprising a permissible explosive covered by a sheath encased in a sealed covering intended to be fired other than in a blasthole.

Shelf Life—The maximum storage period during which an explosive material retains adequate physical characteristics and performance.

Shock Tube—Small diameter plastic tubing containing a small amount of reactive material or a reactive filament on its inside surface. Used for initiating detonators. The reactive material ensures that the detonation shock wave is contained within and guided by the walls of the tube. Can be initiated by special starter equipment or detonators or detonating cord.

Shock Wave—A term used to describe a shock front and its related phenomenon.

Short (Circuit)—An unintended joining together of two or more parts of an electrical circuit.

Short Delay Blasting—A blasting technique involving the delayed firing of multiple charges at intervals measurable in milliseconds.

Shot Anchor—A device used to hold explosive charges in the blasthole so as to prevent their being blown or sucked out by the detonation of other nearby charges.

Shot Firer—(Same as Blaster but used more in underground operations.)

Shunt—1) A commercial device attached to the legwires of electrical detonators by the manufacturers to short circuit it. 2) A deliberate short circuit. 3) The process of twisting together the free ends of detonator legwires to short them out. 4) Shorting any section of an electrical blasting circuit to itself.

Silver Chloride Cell/Battery—A battery of particularly low-current output used in certain blasting galvanometers.

Slurry—An explosive material consisting of substantial amounts of a liquid, oxidizers and fuel, and a thickening agent.

Small-Arms Ammunition—Cartridges intended for use in rifles, pistols, revolvers, propellant-actuated devices, and industrial guns. Not military ammunition containing explosive bursting charges or incendiary, tracer, spotting, or pyrotechnic projectiles.

Small-Arms Ammunition Primers—Small, usually metal housings filled with impact- or stab-sensitive and flame-producing chemicals used to ignite the propellant in, for example, small arms cartridges. Frequently abbreviated to "primer" or "cap".

Smoke—Solid particles suspended in air.

Smokeless Powder/Propellant—Solid propellant used in small-arms ammunition, rockets, propellant-actuated devices, etc.

Snakehole—1) A hole drilled under a boulder. 2) A blasthole drilled downward slightly from the horizontal into the floor elevation of a quarry face.

Socket—(Same as Bootleg.)

Softwood—Douglas fir or a wood of similar bullet resistance without any defect.

Spacing—The distance between blastholes measured (in bench blasting) parallel to the free face and perpendicular to the burden.

Specific Gravity—The ratio of the weight of any volume of substance to the weight of an equal volume of water.

Springing—A technique used to enlarge the bottom of blastholes by firing a small charge. Often used to facilitate the subsequent loading of a larger diameter charge in the same blastholes.

Squib—A pyrotechnic firing device burning with an external flash used to ignite black powder and pellet powder.

Stabilizer—A material or substance added to reduce (sometimes prevent completely) autocatalytic decomposition of an explosive.

Static Electricity—Electricity at rest on an object or person. Often generated by the repeated contact and release of dissimilar insulating materials.

Steady State Velocity—The velocity at which a specific charge diameter of given explosive will detonate.

Stemming—1) Inert material placed in blasthole between or on top of explosive charges for the purpose of keeping them separate or confining them. 2) The process of installing the material.

Stoichiometric—A reactive chemical compound in which the balance of the reactive chemicals is calculated to be such that they all react.

Stray Cat—Electrical current flowing on the outside of an insulated conductor.

Subdrilling—The technique of drilling blasthole below floor level or working elevation level to ensure burden breakage to that level.

Subsonic—Slower than the speed of sound at the elevation in question.

Supersonic—Faster than the speed of sound at the elevation in question.

Sympathetic Detonation—Detonation of explosive material due to a received impulse from another, exploding charge, via air, water, or earth.

Tamping—1) The action of compacting an explosive charge or the stemming material in a blasthole. 2) The stemming material itself.

Tamping Bags—Cylindrical bags of stemming material used in blastholes to confine the explosive charge(s).

Tamping Poles—A nonmetallic pole used to compact charges or stemming material in a blasthole.

Temperature of Ignition—The temperature required to ignite an explosive under specific conditions.

Test Detonator—An IME standard referring to a detonator having 0.40 to 0.45 grams of PETN as a base charge pressed to a specific gravity of 1.4g/cc and primed with standard weights of primer that vary according to the manufacturer.

Thermite—A mix of powdered metal and a metal oxide that provides an intensely hot, gasless reaction.

Toe—A term used in bench blasting referring to excessive burden measured at the floor level of the bench.

Trunkline—(Same as Detonating Cord Trunkline.) The term may also be encountered in the context of certain gas-initiated or shock-tube-initiating systems.

Unconfirmed Detonation Velocity—The speed at which a detonation wave travels through the unconfined explosive.

UL, Inc.—Underwriters Laboratory, Inc. A nationally recognized testing laboratory equipped and qualified to undertake the necessary testing to determine compliance with the appropriate standards and the proper performance of materials and equipment in use.

Volt—The standard unit of electromotive force that is the difference in potential required to make a current of 1 amp flow through a resistance of 1 ohm.

Volume Strength—(Same as Bulk Strength and Cartridge Strength.)

Warning Signal—A visual or audible signal given to alert persons in the vicinity of a blast area prior to the firing.

Waste Acid—Used, spent, or residual acid from a nitration process.

Water Gel—All explosive material consisting of significant amounts of water, fuel, and oxidizers with a cross-linking agent.

Water Resistance—An indication of a material's ability to withstand the effects of water penetration.

Water-Stemming Bags—Water-filled plastic bags incorporating a self-sealing valve. Permissible stemming devices.

Watt—A unit of electrical power, which is 1 joule per second.

Weather Resistant—Offering reasonable protection from the effects of weather.

Weight Strength—The energy of an explosive per unit of weight, which is usually expressed as a percentage of the energy per unit of weight of a specified explosive standard.

INDEX